What They're Saying About *Tornado in a Junkyard*

Dr. Duane T. Gish, Senior Vice President, Institute for Creation Research: "*Tornado in a Junkyard* by James Perloff should be in the library of every one who is interested in the subject of origins. This book is a powerful argument for creation because it is thorough, fully documented, and scientifically accurate. It is easily readable by scientist and layman alike, and is written in a popular style that will make it interesting and entertaining for readers of all ages. I highly recommend this book."

Publisher's Weekly, 8-30-99: "James Perloff's intriguing *Tornado in a Junkyard* aims to debunk evolutionary theory in favor of creationism. Perloff, a former contributing editor to the *New American*, draws upon the work of neo-Darwinists and geneticists to argue that 'while *micro*evolution does occur—meaning minor adaptations and variations within a species,' there is no solid evidence for *macro*evolution, or conversion of one animal type into another."

Dr. Emmett L. Williams, President, Creation Research Society: "*Tornado in a Junkyard* is a unique presentation of the scientific case against Darwinism, informally written for laymen. If you are looking for a user-friendly explanation of the facts supporting creation, this book is for you."

Conservative Book Club, 12-99: "James Perloff brings *all* the data together in a volume readily accessible to nonscientific types. His conclusion, carefully drawn: science contradicts Darwinism. . . . Perloff's style, unusually lively, makes *Tornado in a Junkyard* entertaining as well as educational."

Actor Jack Lemmon, who played Clarence Darrow in the 1999 film version of *Inherit the Wind*: "My congratulations to Mr. Perloff for an outstanding piece of work."

Homeschooling Today, Jan/Feb-2000: "Why another 'anti-evolution' book? Because *Tornado in a Junkyard* is *different*. Author James Perloff, a former fanatical atheist and anti-creationist, understands the other side's point of view. He presents facts that logically disprove Darwinism."

Ellen Myers, Creation Resource Library, Wichita, Kansas: "I've been heavily involved in the creationist movement for many years and am familiar with most of the facts cited in *Tornado*. However, the racy style, the many excellent photos, and especially the less known details and extensive documentation will now make *Tornado* my resource of choice in my work."

The New American, 9-13-99: "Perloff demonstrates—in this reviewer's opinion conclusively— that scientific evidence, when examined honestly, does not support modern Darwinism, but actually contradicts it. . . . This is a very important work, written in an informal and attractive style that is a joy to read."

Vicki Brady, Host, "Homeschooling USA": "With so many books out on the evolution/creation debate it is getting hard to choose from good, better and best. James' book falls in the best category. I recommend that every homeschool family and church have a copy for their libraries."

CSA News (Creation Science Association for Mid-America), Jan-Feb 2000: "This new book may be one of the very best. Easy reading, loaded with great quotes and very well illustrated, this could be an excellent resource book for student papers, or just the right book for Uncle Fred."

Tom Eynon, 30-year career missionary with the Navigators: "I have read many, many books for laymen on creation science over the years, but this is by far the best—easy to read, told with humor. I call it 'creation science for dummies'—meaning people like me."

Massachusetts News, 11-99: "Perloff's book is a powerful synthesis of recent work by microbiologists, physicists and other scientists showing there is no hard evidence for the creation of new species from existing ones."

Janis Hoover, Host, "Cross Talk," WTLN, Orlando, Florida: "Jim's 'product of the sixties' background and layman's approach to technical issues make *Tornado in a Junkyard* extremely readable. I wish I had read this book first when I began my own research into the discrepancies surrounding the theory of evolution."

Christian News, 9-27-99: "*Christian News* highly recommends *Tornado in a Junkyard.*"

Marty Minto, Host, "Straight Talk Live," KPXQ, Phoenix: "*Tornado in a Junkyard* is a must read! It is one of the most informative books on the creation vs. evolution debate."

TORNADO IN A JUNKYARD

THE RELENTLESS MYTH OF DARWINISM

James Perloff

BURLINGTON, MASSACHUSETTS
WWW.JAMESPERLOFF.COM

Also by James Perloff:

The Shadows of Power: The Council on Foreign Relations and the American Decline (1988)

The Case against Darwin: Why the Evidence Should Be Examined (2002)

Truth Is a Lonely Warrior: Unmasking the Forces behind Global Destruction (2013)

First printing, June 1999
Second printing, January 2000
Third printing, July 2000
Fourth printing, March 2001
Fifth printing, November 2002
Sixth printing, October 2003
Seventh printing, August 2006
Eighth printing, October 2010
Ninth printing, May 2016

Printed and bound in the United States of America.

Published by Refuge Books, 25 South Bedford Street, Burlington MA 01803
email: bookrefuge@aol.com website: www.jamesperloff.com

Book design and typography by Arrow Graphics, Inc. info@arrow1.com
Cover illustration by Paul Ingbretson
Cover design by James Bennette

Library of Congress Catalog Card Number: 98-92197

Cataloging-in-Publication Data

Perloff, James
Tornado in a Junkyard: The Relentless Myth of Darwinism/James Perloff—1st ed.
p. cm.
Includes bibliographical references and index.
ISBN-13: 978-0-9668160-0-6
ISBN-10: 0-9668160-0-5
1. Evolution. 2. Creationism. I. Title.
QH366.2.P47 1999
575—dc20 98-92197

Contents

CHAPTER 1

Baby Boomer Tunes Out, Turns On, Bums Out, Burns Out

It was spring, 1963. My sixth-grade teacher, Mr. Halpern, was in the middle of a lesson. "Scientists," he said, "have determined that the Earth is about five billion years old."

Janet Woodward, one of my little classmates, frowned and raised her hand. "But Mr. Halpern, that's not what the Bible says."

He looked at the girl with slight pity. "Well, I'm sorry, but this is something that science has definitely proven. The Bible is simply wrong on that subject."

"But that's not what my parents say," she persisted.

Mr. Halpern, perhaps wanting to head off a complaint from Janet's mother and father, paused, then softened his line. "Well, of course," he said, "you have a right to your religious beliefs, and I'm not saying you shouldn't listen to your parents. But this has been proven scientifically—it's really not open to debate."

I didn't know it then, but this little conversation would be a subtle yet powerful turning point in my life. Certainly, I hadn't grown up religious. My family never prayed or read the Bible. We attended church for a while, but strictly as nominal and indifferent members. The sermons were boring—I don't remember a single thing that was ever said.

Nonetheless, I didn't despise the Bible as a child. My second grade teacher in La Canada, California, always had us read from the Psalms to start the day. We moved to Massachusetts in 1960, and my third grade class put on a nativity play. I was familiar with the Genesis account of how the world was created—"In the beginning," Adam and Eve, that stuff. I didn't have any ideas about it, but I supposed it might be true.

Mr. Halpern, however, shot that down in flames. Teacher was telling us, for a fact, that science had proven the Bible wrong; Adam and Eve were blarney. And if the Old Testament was a fairy tale, the New probably was also. As my teen years dawned, I learned more such ideas. The Bible, I was told, contained some respectable poetry—but beyond that, was valueless. It was a lot of myths and legends, no more valid than Greek mythology. God did not create us; we had evolved from apes, and life began as little cells in the ocean billions of years ago. Man created God. He was like Santa Claus—as a child you might believe in him, and that could be comforting, but later on you wised up and realized there was no such being.

After all, didn't the Bible say men once lived for hundreds of years? And how about that junk about Noah taking all those animals on the Ark? Obviously, bogus. Charlton Heston pulled off miracles in *The Ten Commandments*, but there sure weren't any happening in real life today! In time, I began growing hostile toward religion. After all, ministers committed one crime beyond forgiveness—Saturday morning TV had great cartoons, especially Looney Tunes at 11 A.M. But on Sunday? Nothing but tedious sermons—and in those pre-cable days, you could only pick up three or four stations clearly around Boston.

I noticed something else, too. Some of my nice sixth-grade chums of 1963 were wearing leather jackets and turning into virtual hoodlums in 1964. Of course, everyone knows that teens rebel, and it's rightly chalked up to puberty hormones, that sort of thing. But this was more; a meanness was growing in these kids that didn't seem to be around in the fifties.

In fact, things changed in 1963 on a national level. President Kennedy was assassinated, and the Vietnam War was heating up. *Leave it to Beaver* ran its last episode, and the Beatles were around the corner. America was suddenly different. I didn't know what was going on, but something had undeniably changed.

Later I got interested in liberal politics, especially after watching the 1968 Democratic National Convention on TV. And while in retrospect, perhaps the networks showed only what they wanted, the scenario conveyed clear-cut good guys and bad guys. The good guys were the protesting hippies and yippies. The villains were the Chicago police, Mayor Richard Daley, and the Democratic Party's conservative wing, which wanted to push on in Vietnam. Costa-Gavras's *Z*, a brilliantly produced movie, further consolidated my views.

In the summer of '69, I got to know some hippies from nearby Cambridge. I loved their nonconformity and politics, but especially liked that they accepted me, respected me, and talked with me about adult issues. It was a welcome relief from the juvenile cliquishness of high school society.

That fall, I entered Colby College in Waterville, Maine as a freshman. The nation's battle over Vietnam was heading towards its crescendo and the campus was totally divided. We didn't call ourselves "liberals" and "conservatives" there—we were either "freaks" or "jocks" and I definitely landed in the "freak" camp. A friend and I hitched to D.C. to protest the war. I made my politics vociferously known, and, while drunk, even insulted a running back on the football team. He knocked out one of my teeth, but I didn't press charges; after all, it got me some attention, and besides—I deserved the punch.

I started writing a left-wing column called "Candy Cigarette" for the Colby newspaper, the *Echo*. I always sprinkled in four-letter words, certain that the campus prudes needed a few good jolts. And I wrote a comic strip for the paper as well, spoofing President Richard Nixon and Vice President Spiro Agnew as Batman and Robin. They'd go whizzing about in capes, socking liberals like Kennedyman and McGovernman with big "pows" and "biffs."

The campus conservatives counter-attacked with columns opposing mine. They rotated their forces, but it didn't matter. I was convinced I could run rings around all of them.

After the Kent State shootings in the spring of 1970, Colby joined the rest of the nation's colleges and went on strike. I made a beeline for Cambridge, and rented a room in a hippie house a few blocks from Harvard Square.

Cambridge was a revolutionary fortress, the Haight-Ashbury of the East. Long hair was a badge on a man then, identifying you as "member, rebel fraternity." Strangers would approach you, say "Hi!" and off went the conversation. My roommates and I would yell greetings out the window to other hippies passing in the street; in they'd come, we'd light up joints and rap for hours. Nothing makes friends like mutual enemies—and our bad guys were everywhere: pigs (cops), Nixon, the military-industrial establishment, and anyone who agreed with them. Obviously, the Victorian fogies of the previous generation didn't like our lifestyle because they were jealous. Even celebrities I'd admired as a child—like John Wayne, Bob Hope and Walt Disney—were now on my list of total skuzzballs because of their conservatism.

Cambridge in the summer of '70 was crackling with excitement. Every evening seemed to bring some new adventure. The girls were good-looking, and nearly every song on radio that summer was a total hit. I hawked the then-radical *Boston Phoenix* on the street. I narrowly eluded pigs firing tear gas during a Harvard Square riot. I got mugged one night—and even *that* I dug, moronically chasing the guys through the streets of Cambridge. The next day I read in the paper that the same dudes had been on a crime spree and knifed some people.

I was eighteen and my housemates a few years older, but they treated me as an equal, which I really loved. I shared a room with a hashish dealer. I delivered a few bags of grass for him. Personally, though, I didn't go much for marijuana and hash—couldn't seem to get high off them like other people. So I stuck mostly with booze. But that was cool with everyone—diversity, circa 1970.

When I returned from Colby to Cambridge in the summer of 1971, however, I was shocked at the difference one year had made. No more flower child idealism and brotherhood among long-hairs. The whole movement had grown sullen, with cynicism and infighting. People vied to be cooler than each other. It got to where, if you liked McCartney, you were uncool; only Lennon was cool. The change wasn't just a Cambridge thing, it was national. The music had quickly soured, and hits were becoming few and far between. The Beatles and Supremes had split up. Hendrix, Morrison and Joplin were dead. And that's what happened to the whole hippie thing—divided and dead. It reminded me of 1963-64, when my grade school friends, and the country, had undergone a sweeping transformation. Now, another was mysteriously occurring. I still didn't know why, only that it happened—and since others shared my observation, I knew I wasn't totally bonkers.

By the mid-seventies, Vietnam was winding down, and we rebels were without our biggest cause. The great music was all gone. Rock and roll was eclipsed by disco, which mostly sounded alike to me. In 1970, the number one hit had been the beautiful *Bridge Over Troubled Water*. In 1976: *Disco Duck*. "Boys," I told my roommates, "music has come a long way since the Beatles." We laughed sadly.

One thing didn't change, though—my hatred of religion and the Bible. I was a full-fledged atheist, with only contempt for naive fools who believed. Nothing could bring a smirk to my face like the mention of God or the "m" word—morality. My friends and I, who had *National Lampoon* mentalities, made no end of God jokes. Heaven, we said, was b-o-o-o-ring—strumming a harp all day. But hell? Far-out! There was always a party, an endless supply of booze, and the music was great. On some nights, if you were lucky, you might catch an appearance by the devil himself—a real swell down-to-earth Joe.

The only "judgement day" was going to be the one run by Tom Yawkey, the Red Sox owner who died in 1976. Yawkey had invested millions in high salaries for over forty years—and the players never won him a world championship. On judgement day, Yawkey would gather before him all the goats, each of whom would receive punishment fitting his deeds. The players would plead for mercy, but Uncle Tom would read off the batting averages and winning percentages, and their fates would be sealed. Johnny Pesky, who made that wild throw to blow the '46 World Series?—he'd get endless mouthfuls of baseballs. "Taste good, Pesky?" Uncle Tom would ask with a chuckle. Don Buddin—the shortstop who booted all those grounders back in the fifties? He'd ride down a bannister that turned into a razor blade. Bill Buckner? Well, we didn't know about him yet.

That was the mild stuff. One friend and I invented jokes about God and Jesus Christ that were so crudely obscene that I doubt the blasphemy has ever been greatly exceeded.

Very rarely, someone would try to approach me about God. I hitchhiked a lot in the early seventies, and an ex-heroin addict who gave me a ride witnessed about Jesus Christ. Well, I didn't really mind, he was an OK guy. Then an older Afro-American picked me up hitching and talked about God. I smirked, nodded, and thanked him for the lift. I couldn't really talk back to these people—after all, they were doing me a favor. (Then again, it did seem like a ripoff to give people rides just to push your religion on them!)

But if you came to me about God from out of the blue—watch out! On one job where I worked, a girl just mentioned God in passing. "God?" I said. "*God*? Come on! Get real, will ya? Everyone knows there isn't any God! I mean, where have you been? What do you think this is, the Middle Ages?" Her eyes bristled. Well, that was her problem. Next time she'd know to keep her stupid yap shut.

My animosity toward religion was so great that I even studied the play *Inherit the Wind* to see what proofs against God I could learn.

However, my own life was beginning to sag. Boozing it up and listening to "Born to be Wild" through stereo headphones can only go on for so long. I was getting older, but not better. And if I'd been frank with myself, I would have admitted that underneath the surly surface, I wasn't happy. I wasn't competent at too many things; I was socially inept and frequently dishonest.

Some people approached me with a New Age perspective. Initially, I rebuffed them. After all, I was against religion, not just Christianity. But with enough persuasion and self-dissatisfaction, I cast off animosity to all things spiritual, and began a long trek into the mysteries of the New Age movement, pursuing the god I could become, only to find myself empty after a decade of fruitless searching.

What's the point? Oh, no, is this one of them sappy conversion stories? Mr. burnt-out hippie couldn't cut it any longer, so he got religion? Oh, break out the violin and hankies! Spare us!

Actually, I'm done talking about myself. Truthfully, I needed to establish that this book wasn't written by someone raised as a fundamentalist. Now, let's turn the clock back, back, past Reagan, past disco, past Woodstock, back to 1963, to that moment when Mr. Halpern told my sixth-grade class that science had proven the Earth was five billion years old and so the Bible was all wrong—a moment that started me down a road a lot of you traveled with me . . .

CHAPTER 2

Problems Carved in Stone

This book deals with science and, to a lesser extent, religion—not exactly topics most of us make beelines for at the bookstore. Wouldn't we all rather enjoy some *Dilbert* and *Far Side* cartoons? However, science and religion are significant subjects. Ultimately, they address who we are, why we are, where we came from, and where we're going.

The prevailing explanation for life runs something like this. About fifteen billion years ago, all the universe's matter and energy were compressed together, but for some reason exploded in a big bang. This was a hot process, but as things cooled, hydrogen and helium gas formed. The gas molecules collapsed on themselves to make stars and galaxies. Thus the universe came to be.

Our solar system formed five billion years ago from a cloud of dust and gas, which condensed into the sun and planets. (In a popular older theory, the planets spun off from the sun.) A long time ago, Earth was molten rock. As its heat dissipated, oceans of warm primordial "soup" presented ideal conditions for the origin of life.

About three billion years ago, life began as simple cells. Eventually, these evolved into multicellular organisms, which then became invertebrates (creatures with no backbones, like jellyfish and clams). These in turn evolved into vertebrates and the first fish.

After a while, some fish tired of the water and yearned to go on dry land. Over eons, as fish struggled to get ashore, they developed little legs and finally succeeded, becoming amphibians (frogs, salamanders, etc.). Amphibians then evolved into reptiles and reptiles into mammals and birds.

The process by which this occurred is called "natural selection." Organisms best suited to their environment were able to survive, and passed their strengths on to the next generation. Thus, over time, creatures became more highly developed—in much the same way, we are told, that breeders develop better fruits, flowers or racehorses. Less fit creatures became extinct.

Each organism resulted from adaptations to its environmental niche. Giraffes, for example, lived around high vegetation. Since giraffes with longer necks survived better, long necks became standard in the species. In the tooth and claw struggle for existence, each animal has evolved survival mechanisms. We see this illustrated everywhere: chameleons can change colors and so conceal themselves from predators; porcupines have needles to fend off enemies, and so forth.

Man, it is said, descended from ape-like creatures. As he evolved and became more intelligent, his brain and skull grew larger. He formerly swung from trees, but, after adapting to life on the ground, he lost his tail as it no longer served any purpose. Because mates with less hair were more attractive, man eventually lost his ape-like hair as well. Later he reached the caveman stage: still a brute, but able to use crude stone tools. Finally, he evolved to his modern state.

That, more or less, is the explanation for life and the universe that schools teach today. Until the nineteenth century, the Biblical view—that God had created the world and man—was almost universal in the West. But after publication of Darwin's *The Origin of Species* in 1859, evolutionary ideas began replacing religious orthodoxy, until evolution itself became orthodoxy.

In recent years, however, Darwinism has been challenged—not on religious grounds, but compelling scientific ones. Books such as *Evolution: A Theory in Crisis* by molecular biologist Michael Denton, *Darwin's Black Box* by biochemist Michael Behe, and *Darwin on Trial* by law professor Phillip Johnson, to name just a few, have shaken the evolutionary establishment.

This book will examine the growing case against evolution. That means we'll deal with—ugh—"science stuff." But we'll try to keep it simple and interesting, like those *Dummies* books. We'll define scientific words, even at the risk of annoying those well-versed in such terminology.

Speaking of definitions, what is science? *Webster's New World Dictionary* describes it as "systematized knowledge derived from observation, study and experimentation."

Science depends on observations, not subjective opinions. What observations, then, support the theory of evolution? How can we know that fish evolved into land creatures and reptiles into birds, especially since this happened millions of years ago, before we were around to see it?

The only real way to know the past is to consult records—in this case, the fossil record. I quote Pierre-Paul Grassé, the most eminent French zoologist of his day:

> Naturalists must remember that the process of evolution is revealed only through fossil forms. A knowledge of paleontology [the study of fossils] is, therefore, a prerequisite; only paleontology can provide them with the evidence of evolution and reveal its course or mechanisms.[1]

Yale paleontologist Carl Dunbar, a stout defender of evolution, agreed:

> Although the comparative study of living animals and plants may give very convincing circumstantial evidence, fossils provide the only historical, documentary evidence that life has evolved from simpler to more and more complex forms.[2]

Sir Gavin de Beer, director of the British Museum, affirmed: "The last word on the credibility and course of evolution lies with the paleontologists. . . ."[3]

Fossils are impressions or remains of plant and animal life preserved in the earth. They are found in different media such as sedimentary rocks, volcanic ash, coal or amber. A bone or shell can constitute a fossil; so can a footprint.

Of the 329 families of land vertebrates living today, 79 percent are represented in the fossil record—88 percent excluding birds (which don't readily fossilize).[4] There are millions of fossils in museums, representing some 250,000 species.[5] National polls usually consider just 2,000 people enough to accurately sample our entire nation. So we certainly have enough fossil specimens to verify Darwinism.

This is especially true since evolution is a gradual process, requiring millions of years. Darwin stated that "the number of intermediate and transitional links, between all living and extinct species, must have been inconceivably great. But assuredly, if this theory be true, such have lived upon the earth."[6] Thus, the fossil record should depict evolution's history: organisms progressing through their stages of development.

As mentioned, Darwinism claims fish transformed into land creatures by evolving little arms and legs over eons. If true, there should be innumerable fossils of fish with rudimentary arms and legs. Yet we do not find them! In fact, *all* organisms appear in the fossil record fully formed, without transitional stages.

Darwin himself recognized this problem. He noted in *The Origin of Species*:

> Why then is not every geological formation and every stratum full of such intermediate links? Geology assuredly does not reveal any such finely-graduated organic chain; and this, perhaps, is the most obvious and serious objection which can be urged against the theory. The explanation lies, as I believe, in the extreme imperfection of the geological record.[7]

Interestingly, most of the nineteenth-century criticism of Darwinism came not from clergymen, but scientists. French paleontologist François Jules Pictet complained:

> Why don't we find these gradations in the fossil record, and why, instead of collecting thousands of identical individuals, do we not find more intermediary forms? To this Mr. Darwin replies that we have knowledge of such a small proportion of fossils that one cannot construct proofs. . . . Consequently, according to him, we have only a few incomplete pages to the great book of nature and the transitions have been in pages which we lack. By why then and by what peculiar rules of probability does it happen that a species which we find most frequently and most abundantly in all the newly discovered beds are in the immense majority of the cases species which we already have in our collections?[8]

Darwin hoped more time and excavations would yield fossils supporting his theory. He explained that "Only a small portion of the surface of the earth has been geologically explored and no part with sufficient care"[9] and "We continually forget how large the world is, compared with the area over which our geologic formations have been carefully examined."[10]

But in Darwin's lifetime, nothing improved, and he lamented: "When we descend to details, we can prove that no one species has changed (i.e., we cannot prove that a single species has changed)."[11]

Are things different now? Anthropologist Edmund R. Leach told the 1981 Annual Meeting of the British Association for the Advancement of Science: "Missing links in the sequence of fossil evidence were a worry to Darwin. He felt sure they would eventually turn up, but they are still missing and seem likely to remain so."[12]

Let's hear from David Raup, curator of geology at the Field Museum of Natural History in Chicago, which houses the world's largest fossil collection:

> He [Darwin] was embarrassed by the fossil record because it didn't look the way he predicted it would and, as a result, he devoted a long section of his *Origin of Species* to an attempt to explain and rationalize the differences. . . . Darwin's general solution to the incompatibility of fossil

evidence and his theory was to say that the fossil record is a very incomplete one. . . . Well, we are now about 120 years after Darwin, and knowledge of the fossil record has been greatly expanded. We now have a quarter of a million fossil species but the situation hasn't changed much. The record of evolution is still surprisingly jerky and, ironically, we have even fewer examples of evolutionary transition than we had in Darwin's time. By this I mean that some of the classic cases of Darwinian change in the fossil record, such as the evolution of the horse in North America, have had to be discarded or modified as a result of more detailed information—what appeared to be a nice simple progression when relatively few data were available now appears to be much more complex and much less gradualistic.[13]

Harvard paleontologist Stephen Jay Gould, probably evolution's leading spokesperson today, has acknowledged:

The extreme rarity of transitional forms in the fossil record persists as the trade secret of paleontology. The evolutionary trees that adorn our textbooks have data only at the tips and nodes of their branches; the rest is inference, however reasonable, not the evidence of fossils.[14]

Colin Patterson, senior paleontologist at the British Museum of Natural History, wrote in a personal letter:

I fully agree with your comments on the lack of direct illustration of evolutionary transitions in my book. If I knew of any, fossil or living, I would certainly have included them. You suggest that an artist should be used to visualize such transformations, but where would he get the information from? I could not, honestly, provide it, and if I were to leave it to artistic license, would that not mislead the reader?

I wrote the text of my book four years ago. If I were to write it now, I think the book would be rather different. Gradualism is a concept I believe in, not just because of Darwin's authority, but because my understanding of genetics seems to demand it. Yet Gould and the American Museum people are hard to contradict when they say there are no transitional fossils. As a paleontologist myself, I am much occupied with the philosophical problems of identifying ancestral forms in the fossil record. You say that I should at least "show a photo of the fossil from which each type of organism was derived." I will lay it on the line—there is not one such fossil for which one could make a watertight argument.[15]

George Gaylord Simpson, perhaps the twentieth century's foremost paleontologist, said:

This regular absence of transitional forms is not confined to mammals, but is an almost universal phenomenon, as has long been noted by paleon-

> tologists. It is true of almost all orders of all classes of animals, both vertebrate and invertebrate.[16]

David B. Kitts of the School of Geology and Geophysics at the University of Oklahoma wrote:

> Despite the bright promise that paleontology provides a means of "seeing" evolution, it has presented some nasty difficulties for evolutionists, the most notorious of which is the presence of "gaps" in the fossil record. Evolution requires intermediate forms between species and paleontology does not provide them.[17]

Dr. Steven Stanley of the Department of Earth and Planetary Sciences, Johns Hopkins University, adds: "The known fossil record fails to document a single example of phyletic evolution accomplishing a major morphologic [structural] transition and hence offers no evidence that the gradualistic model can be valid."[18]

Professor Heribert Nilsson, director of the Botanical Institute at Lund University, Sweden, declared after forty years of study: "It may, therefore, be firmly maintained that it is not even possible to make a caricature of an evolution out of paleobiological facts. The fossil material is now so complete that it has been possible to construct new classes and the lack of transitional series cannot be explained as due to the scarcity of the material. The deficiencies are real, they will never be filled."[19] He complained:

> If a postulated ancestral type is not found, it is simply stated that it has not so far been found. Darwin himself often used this argument and in his time it was perhaps justifiable. But it has lost its value through the immense advances of paleobiology in the twentieth century. . . . The true situation is that those fossils have not been found which were expected. Just where new branches are supposed to fork off from the main stem it has been impossible to find the connecting types.[20]

Gareth J. Nelson of the American Museum of Natural History:

> It is a mistake to believe that even one fossil species or fossil "group" can be demonstrated to have been ancestral to another. The ancestor-descendant relationship may only be assumed to have existed in the absence of evidence indicating otherwise.[21]

To quote a more familiar voice, *Newsweek* reported in 1980:

> In the fossil record, missing links are the rule: the story of life is as disjointed as a silent newsreel, in which species succeed one another as abruptly as Balkan prime ministers. The more scientists have searched for the transitional forms between species, the more they have been frustrated.[22]

By citing so many authorities, I haven't meant to beat a dead horse—only to demonstrate beyond question that the fossil record does not support evolution. This is true for every class of animal.

Supposedly, microscopic organisms evolved into the first invertebrates. Today we have innumerable fossils from the Earth's most ancient rocks—the "Cambrian" and (much more rarely) "Precambrian" periods. Yet there are no transitional fossils linking microorganisms and complex invertebrates.

You have probably heard of the "Cambrian explosion." Cambrian rock contains numerous marine organisms which appeared suddenly—their fossil ancestors completely missing. Stephen Jay Gould acknowledges that "our more extensive labor has still failed to identify any creature that might serve as a plausible immediate ancestor for the Cambrian faunas."[23] British biologist Richard Dawkins, an ardent evolutionist, notes:

> [T]he Cambrian strata of rocks, vintage about 600 million years, are the oldest ones in which we find most of the major invertebrate groups. And we find many of them already in an advanced state of evolution, the very first time they appear. It is as though they were just planted there, without any evolutionary history.[24]

Supposedly, invertebrates evolved into vertebrates—surely a very long process. Yet despite countless fossils from both groups, there is not one specimen intermediate between them! If such existed, museums would display them.

Where did the first fish come from? There are billions of fish fossils, but as J. R. Norman of the British Museum of Natural History stated: "The geological record has so far provided no evidence as to the origin of the fishes. . . ."[25] Gerald T. Todd, in *American Zoologist*, elaborated:

> All three subdivisions of the bony fishes first appear in the fossil record at approximately the same time. They are already widely divergent morphologically, and they are heavily armored. How did they originate? What allowed them to diverge so widely? How did they all come to have heavy armor? And why is there no trace of earlier, intermediate forms?[26]

The next group, amphibians, also appears suddenly in the fossil record. For years, evolutionists claimed that the coelacanth, a bony fossil fish, was a forerunner of the amphibians, its fins described as limb-like. The coelacanth had supposedly been extinct for 70 million years.

Then, in 1938, fishermen caught a live one off the African coast. Since then, about 200 more have been caught. This embarrassed evolutionists, for besides proving the coelacanth was not extinct for 70 million years, examination revealed it was 100 percent fish, with no amphibian characteristics.

Amphibians supposedly evolved into reptiles. But as paleontologist Robert L. Carroll noted in *Biological Reviews of the Cambridge Philosophical Society*:

> Unfortunately not a single specimen of an appropriate reptilian ancestor is known prior to the appearance of true reptiles. The absence of such ancestral forms leaves many problems of the amphibian-reptile transition unanswered.[27]

From Barney to *Jurassic Park*, dinosaurs have become a hot item. But how many dinosaur ancestors occur in the fossil record? *Not one, even though they supposedly roamed Earth for 165 million years*. Some dinosaurs had spikes, some armor plates. Their contemporaries, the pterodactyls, had wings. Where are all the transitional fossils showing how these features developed?

This problem among prehistoric reptiles was recognized even in the nineteenth century. Sir Richard Owen, who coined the word "dinosaur," and superintended the Department of Natural History at the British Museum, noted:

> The last ichthyosaurus, by which the genus disappears in chalk, is hardly distinguishable specifically from the first ichthyosaurus, which abruptly introduces that strange form of sea-lizard in the Lias. The oldest Pterodactyle is as thorough and complete a one as the latest.[28]

Reptiles allegedly became mammals. There are fossils that have been labeled "mammal-like reptiles," but since they are all extinct, we cannot examine their soft tissues. As molecular biologist Michael Denton observes:

> The possibility that the mammal-like reptiles were completely reptilian in terms of their anatomy and physiology cannot be excluded. The only evidence we have regarding their soft biology is their cranial endocasts and these suggest that, as far as their central nervous systems were concerned, they were entirely reptilian.[29]

And even these lack any direct fossil intermediates. Tom Kemp, curator of zoology at the Oxford University Museum, wrote:

> Gaps at a lower taxonomic level, species and genera, are practically universal in the fossil record of the mammal-like reptiles. In no single adequately documented case is it possible to trace a transition, species by species, from one genus to another.[30]

To account for the oddity of oceanic mammals (whales, sea cows and dolphins), evolutionists say they evolved from fish to land creatures, then went back to the sea again. But British author Douglas Dewar, a fellow of the Zoological Society, noted:

> Both whales and sea cows swim by the up and down movement of the great flattened tail. Such movement is impossible in a land animal that has a pelvis, but a well-developed pelvis is essential to every land animal which uses its hind legs for walking. . . . I have repeatedly asked evolutionists to describe or draw the skeleton of a creature of which the pelvis and hind legs are anatomically midway between the state that prevails in whales and sea cows on the one hand, and a land quadruped on the other. No one has accepted the challenge, and of course a fossil of such a creature has not been found, and never will be.[31]

Reptiles supposedly became birds as well, their scales turning into feathers. But what documentation exists? W. E. Swinton of the British Museum of Natural History stated: "The origin of birds is largely a matter of deduction. There is no fossil evidence of the stages through which the remarkable change from reptile to bird was achieved."[32] Barbara J. Stahl, in *Vertebrate History: Problems in Evolution*, wrote: "No fossil structure transitional between scale and feather is known, and recent investigators are unwilling to found a theory on pure speculation."[33]

What about insects? Peter Farb, author of the Life Nature Library's book *The Insects*, observed:

> There are no fossils known that show what the primitive ancestral insects looked like. . . . Animals of such complexity do not come into being suddenly; they must have been evolving for tens of millions of years before. Until fossils of these ancestors are discovered, however, the early history of the insects can only be inferred.[34]

Likewise, plants lack fossil evidence for their evolution. They have no ancestral forms in the geologic layers. Professor Chester A. Arnold, University of Michigan, noted in *An Introduction to Paleobotany*:

> It has long been hoped that extinct plants will ultimately reveal some of the stages through which existing groups have passed during the course of their development, but it must be freely admitted that this aspiration has been fulfilled to a very slight extent, even though paleobotanical research has been in progress for more than one hundred years. As yet we have not been able to trace the phylogenetic history of a single group of modern plants from its beginning to the present.[35]

Supposedly, marine plants evolved into terrestrial plants, but as Gensel and Andrews reported in *American Scientist*:

> We still lack any precise information concerning the presumed aquatic ancestors from which land plants evolved, and the search for evidence of these precursors and of probable transitional stages continues.[36]

E. J. H. Corner, professor of tropical botany at Cambridge University, though opining that other fields support evolution, stated that:

> . . . I still think that, to the unprejudiced, the fossil record of plants is in favour of special creation. If, however, another explanation could be found for this hierarchy of classification, it would be the knell of the theory of evolution. Can you imagine how an orchid, a duckweed, and a palm have come from the same ancestry, and have we any evidence for this assumption? The evolutionist must be prepared with an answer, but I think that most would break down before an inquisition. Textbooks hoodwink.[37]

Some will ask: "What about the horse? Isn't that a well-documented example of evolution?" The famous "horse sequence" originated with Othniel C. Marsh, a nineteenth-century fossil hunter who dug up dozens of horse fossils. Marsh put them in a logical-looking arrangement that Yale University displayed. The series was long claimed to demonstrate progression from smaller, multi-toed horses to the larger, one-toed horse of today.

However, the sequence is discredited. *Eohippus*, the tiny creature once considered the first horse, is now known to more closely resemble the modern hyrax, a rodent-like mammal. It is true that, in North America, one-toed horses are found in more recent geologic layers than three-toed. However, excavations in Nebraska have shown that three-toed and one-toed horses coexisted.[38] And in South America, a one-toed horselike creature, *Thoatherium*, became extinct before three-toed types of the same order did.[39]

It is quite easy to arrange some fossils from smallest to largest, and claim this proves evolution. One could, for example, place the bones of a Chihuahua, beagle, boxer, and Great Dane in sequence, then tell a future generation that they show dog evolution. This is essentially what happened with the horse. Even today, however, horses vary greatly, from miniatures and ponies up to work horses.

Lest anyone think me impertinent in calling the horse sequence dead, I have already quoted (p. 11) David Raup of Chicago's Field Museum. But let's hear additional authorities.

Professor Heribert Nilsson of Lund University:

> The family tree of the horse is beautiful and continuous only in the textbooks. In the reality provided by the results of research it is put together from three parts, of which only the last can be described as including

> horses. The forms of the first part are just as much little horses as the present-day damans [hyraxes] are horses.[40]

Garrett Hardin, professor of biology at the University of California, Santa Barbara:

> [T]here was a time when the existing fossils of the horses seemed to indicate a straight-line evolution from small to large, from dog-like to horse-like, from animals with simple grinding teeth to animals with the complicated cusps of the modern horse. It looked straight-line—like the links of a chain. But not for long. As more fossils were uncovered, the chain splayed out into the usual phylogenetic [evolutionary] net, and it was all too apparent that evolution had not been in a straight line at all, but that (to consider size only) horses had now grown taller, now shorter, with the passage of time. Unfortunately, before the picture was completely clear, an exhibit of horses as an example of orthogenesis had been set up at the American Museum of Natural History, photographed, and much reproduced in elementary textbooks (where it is still being reprinted today).[41]

Boyce Rensberger, senior editor of *Science 80*:

> The popularly told example of horse evolution, suggesting a gradual sequence of changes from four-toed, fox-sized creatures, living nearly 50 million years ago, to today's much larger one-toed horse, has long been known to be wrong. Instead of gradual change, fossils of each intermediate species appear fully distinct, persist unchanged, and then become extinct. Transitional forms are unknown.[42]

Another popularly cited fossil is the extinct bird *Archaeopteryx*, whose name means "ancient wing." Found in the same geologic strata as dinosaurs, it was considered ancestral to modern birds, predating them by 60 million years. Evolutionists have long called it intermediate between reptiles and birds. *Archaeopteryx* was noted to have feathers like a bird—but, like a reptile, teeth, claws and a rather long tail.

That exemplifies something evolutionists often do: equate similarities to relationships. Thoughtful consideration and new discoveries have demonstrated that *Archaeopteryx* was a true bird. Only birds have feathers. *Archaeopteryx* had perfectly developed wings; as Stahl notes, "its feathers gave no hint of primitive features. The imprint they left in the rock, clear and sharp, makes it evident that the feathers of *Archaeopteryx* were already in Jurassic time exactly like those of birds flying today."[43] It also had a beak, bird-like skull, perching feet, and wishbone—all unique to birds. In *Archaeopteryx* fossils, none of these traits are in a transitional stage, but fully developed.

As for its "reptile" characteristics, yes, it had claws on its wings, but so does the ostrich, and nobody considers it part reptile. True, *Archaeopteryx* had teeth, but so did some other fossil birds, and its teeth differed distinctly from those of reptiles.[44] Furthermore, some reptiles, such as turtles, have no teeth—they aren't a distinct reptilian trademark. As to *Archaeopteryx*'s tail, further inspection has shown it strongly resembles a swan's.[45]

A fossil resembling a modern bird has now been found in Eastern Colorado in the same geologic strata as *Archaeopteryx*,[46] and Texas Tech researchers have reported discovering bird fossils in rocks dated much older.[47] This eliminates *Archaeopteryx* as the ancestor of today's birds. Even leading evolutionists Stephen Jay Gould and Niles Eldredge discount it as a transitional form, terming it nothing more than a "curious mosaic."[48]

Die-hard evolutionists continue to assert that intermediates exist, but since they must continually cite the same examples—the horse, *Archaeopteryx*, mammal-like reptiles, and scant others—we see just how rare they are. (Of course, claims are also made regarding *human* transitional forms—australopithecines, Neanderthals, etc. This subject is so vast, we will later address it separately.)

To rephrase the transitional problem: Throughout the geologic layers—which supposedly formed over eons—the various kinds of fossils remain essentially unchanged in appearance. They show no evolution over long ages. Paleontologists call this "stasis." *Science 80* editor Boyce Rensberger, reporting comments by Niles Eldredge, noted: "Species simply appear at a given point in geologic time, persist largely unchanged for a few million years, and then disappear."[49] Tom Kemp stated in *New Scientist*:

> As is now well known, most fossil species appear instantaneously in the fossil record, persist for some millions of years virtually unchanged, only to disappear abruptly. . . .[50]

As Harvard's Stephen Jay Gould put it:

> Most species exhibit no directional change during their tenure on earth. They appear in the fossil record looking much the same as when they disappear; morphological change is usually limited and directionless.[51]

Bees are a perfect example of stasis. Their fossils are much like today's bees. In 1988, *Geotimes* commented on the discovery of the oldest bee ever found in amber:

> David Grimaldi, a curator at the American Museum of Natural History, N.Y., reported in December that the stingless bee's advanced features show that bees have changed little in the last 80 million years.[52]

Likewise, arthropods have always been arthropods, clams clams, and fish fish. Sponges have always been sponges, jellyfish jellyfish, and snails snails. Creation scientist Duane Gish commented with irony:

> Thus evolutionists would have us believe that while some chordate evolved into a fish, which evolved into amphibians, which evolved into reptiles, which evolved into birds and mammals, and lower mammals evolved on up the ladder to humans, all under compelling changes in the environment, chordates have remained unchanged for at least 600 million years! Evolution is a strange phenomenon indeed.[53]

Amphibians remain generally unchanged in the fossil record; so do mammals. For example, paleontologist Robert L. Carroll of McGill University wrote: "The oldest skeleton of a bat, *Icaronycteris* from the early Eocene [about 50 million years ago in the evolutionary time frame], appears almost indistinguishable from living bats."[54]

Fossil plants demonstrate stasis, too. Dr. Gary Parker, a biologist who renounced evolution, asks:

> Did you ever wonder what kind of plants the dinosaurs tromped around on? The answer may surprise you. These unfamiliar animals, now extinct, wandered around among some very familiar plants: oak, willow, magnolia, sassafras, palms, and other such common flowering plants.[55]

If, as Darwinism proposes, bacteria evolved all the way into man, then adaptive changes should have been constantly ongoing. Why, then, do fossils stay the same through countless millions of years?

Would it be worth pointing out that the fossil record, showing all animals complete when first seen, is what we'd expect if God created them whole, just as the Bible says?

PLATE 1. Coelacanth fossil in Jurassic rock

PLATE 2. The coelacanth was thought extinct for 70 million years until caught off Madagascar.

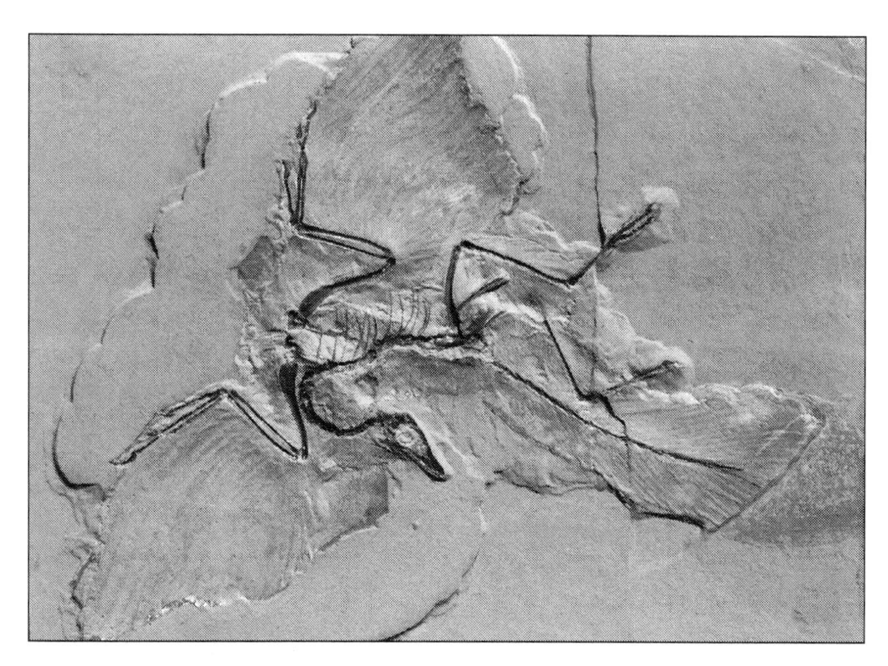

PLATE 3. Fossil of *Archaeopteryx*, purported link between reptiles and birds

PLATE 4. Artist's restoration of *Archaeopteryx*

PLATE 5. Insects preserved in amber demonstrate stasis.

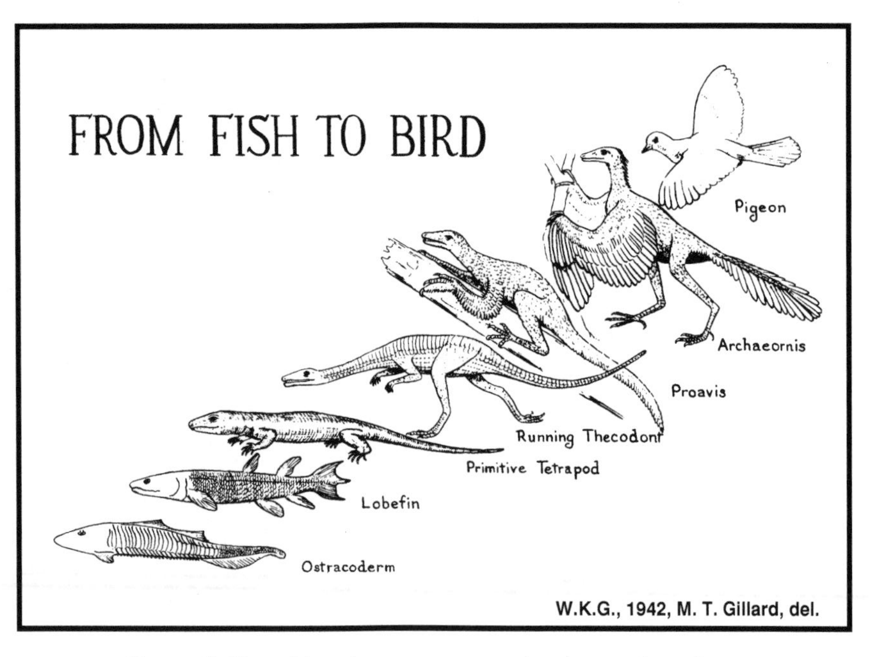

PLATE 6. Transitional sequences: easier drawn than done

CHAPTER 3

Marvelous Mutations

Before continuing, I should mention that using scientific quotations in the creation-evolution debate has stirred controversy. Some feel, quite heatedly, that the opinions of any creation scientist (a scientist who believes the Bible accurately explains creation) carry no validity. For example, Iowa State University professor John Patterson stated in the *Journal of the National Center for Science Education*: "As a matter-of-fact, creationism should be discriminated against. . . . No advocate of such propaganda should be trusted to teach science classes or administer science programs anywhere or under any circumstances."[1]

I believe that a person's religious convictions do not invalidate his qualifications as a scientist. Isaac Newton and Louis Pasteur were devout Christians who trusted the Bible and believed in divine creation. Should we therefore expunge their discoveries? A scientific observation stands or falls on its own merits, not the observer's spiritual outlook.

Still, all but a few of my quotes come from evolutionists. Such usage has also been controversial, however, because it is said that quoting an evolutionist in support of creation misrepresents the person. However, much dissent exists within the evolutionary community. And where a Darwinist comments, for example, on the fossil record's failure to document transitional forms, it is valid to quote him in that context. Let me be clear that in citing evolutionists on specific topics, I am not implying they are creationists. Most would strongly reject this book's broader conclusions.

But the critics can't have it both ways—if all quotes by creation scientists are "invalid," and all quotes from evolutionary scientists are "misrepresentations," we get a two-headed coin that permits no references at all. I will therefore quote both freely, based only on the academic credentials of the individuals, and the apparent merit of their observations.

Except where especially relevant, I will not slow the text by trying to annotate each person's religious beliefs; in most cases I would have no idea what they were. However, everyone I quote who has publicly taken a position as a creation scientist is listed in Appendix One of this book. Those who consider religious beliefs a yardstick of scientific reliability are welcome to consult it.

* * *

Besides a near-total lack of fossil evidence, another problem confronts Darwin's thesis: fish could not develop those little arms and legs unless they first had *genes* for arms and legs. Where'd the genes come from?

In living cells, hereditary information is carried by threadlike structures called chromosomes; genes are the small portions of chromosomes concerned with specific characteristics. Of course, people are born with red hair and brown eyes only if they have genes for red hair and brown eyes. Genes dictate an organism's entire physical makeup, to the minutest detail.

The nineteenth-century monk Gregor Mendel laid the groundwork for genetics in his now-famous botanical experiments. He was ignored in his own time, however, and not until the twentieth century was genetics established as a science. The field was unknown to Darwin, who simply thought creatures adapted to environments, unaware that no change occurs without appropriate genes.

The discovery of genetics thrust a brick wall before evolution. How does a microbe (single cell) turn into a clam, into a fish, into a frog, into a lizard, into an ape, into a man? These transformations couldn't happen unless the organism first had the innumerable genes required for them.

Modern evolutionists came up with an answer: *mutations*. Mutations, of course, are abrupt alterations in genes or chromosomes. Estimates of mutation frequency vary (it depends on method of calculation as well as the type of organism being considered), but everyone agrees they occur quite rarely. Darwinist Richard Dawkins notes that "the probability that a gene will mutate is often less than one in a million."[2] Fritjof Capra, director of the Center for Ecoliteracy in Berkeley, California, reported in 1996: "It has been estimated that those chance errors occur at a rate of about one per several hundred million cells in each generation."[3] Mutations can result from intense causes such as radiation, may produce physical changes, and are sometimes transmitted to offspring.

To make genetics jive with Darwinism, modern evolutionists say natural selection works harmoniously with mutations. In other words, an organism develops some new positive characteristic through mutation, better adapting it to the environment. The creature then passes this mutated trait on to the next generation, while organisms without it, being weaker, die out. A perfect example, we are told, is mutated bacteria which resist antibiotics such as penicillin, thus creating a new strain of stronger bacteria.

Evolution postulates that mutations produced all the changes that have brought the microbe up the ladder to a man. Indeed, they *must* have for Darwinian theory to work; if life began as a single cell, then the entire living world around us resulted from mutations.

This explanation faces serious difficulties, however. Mutations are almost universally *harmful*. In human beings, they are classed as "birth defects." They often result in death or sterility. People today suffer from more than 4,000 disorders caused by gene mutations.[4] Down's syndrome, cystic fibrosis and sickle cell anemia are familiar examples. But what benefits can we list from mutation? Who hopes his child will be born with one?

The human heart is an ingenious structure. Blood is pumped from the right side of the heart to the lungs, where it receives oxygen; back to the heart's left side, which propels it to the rest of the body through more than 60,000 miles of vessels. The heart has four chambers; a system of valves prevents backflow into any of these; electrical impulses from a pacemaker control the heart's rhythm.

Rarely, babies are born with congenital heart disorders, making blood shunt to the wrong place. There is no known case of mutations improving circulation. Hemoglobin—the blood's oxygen-carrying component—has over forty mutant variants. Not one transports oxygen as well as normal hemoglobin.[5]

To accept evolution, we must believe that human blood circulation—a wonder of engineering—was actually *constructed* by chance mutations, when actual observation demonstrates they do nothing but damage it. We must believe that mutations built the human brain and every other feature of life on Earth. British science writer Francis Hitching noted:

> Two of the most powerful causes of mutation are mustard gas and x-rays. A moment's reflection on the horror of Hiroshima children born with deformed limbs and bodies, or blood disorders condemning them to premature deaths, is enough to show that they were unlikely candidates, to say the least, to win the struggle for existence in a life-game where survival of the fittest is the governing rule.[6]

Biochemist Ernst Chain, who shared a Nobel Prize for his work on penicillin, declared:

> To postulate, as the positivists of the end of the last century and their followers here have done, that the development and survival of the fittest is *entirely* a consequence of chance mutations, or even that nature carries out experiments by trial and error through mutations in order to create liv-

> ing systems better fitted to survive, seems to me a hypothesis based on no evidence and irreconcilable with the facts. . . . These classical evolutionary theories are a gross oversimplification of an immensely complex and intricate mass of facts, and it amazes me that they were swallowed so uncritically and readily, and for such a long time, by so many scientists without a murmur of protest.[7]

Even Theodosius Dobzhansky, one of the twentieth century's leading Darwinists, acknowledged:

> And yet, a majority of mutations, both those arising in laboratories and those stored in natural populations, produce deteriorations of viability, hereditary diseases, and monstrosities. Such changes, it would seem, can hardly serve as evolutionary building blocks.[8]

Natural selection does not tend to preserve mutations; rather, it weeds them out. I again quote Mr. Hitching:

> On the face of it, then, the prime function of the genetic system would seem to be to resist change: to perpetuate the species in a minimally adapted form in response to altered conditions, and if at all possible to get things back to normal. The role of natural selection is usually a negative one: to destroy the few mutant individuals that threaten the stability of the species.[9]

The theory has yet another gap. For mutations to occur, there must first be genes. But how did genes originate before the first mutation?

Let's look closer at what mutations really are: informational errors. The genetic code presents efficient instructions for all functions of the human body, or any other organism. Mutations are chance events which alter those instructions. They create a loss of information—not a gain. Dr. Lee Spetner holds a Ph.D. in physics from MIT; he taught information and communication theory for ten years at Johns Hopkins University and the Weizman Institute. His 1997 book *Not By Chance! Shattering the Modern Theory of Evolution* examines mutational theory in detail. Spetner notes:

> But in all the reading I've done in the life-sciences literature, I've never found a mutation that added information. . . . All point mutations that have been studied on the molecular level turn out to reduce the genetic information and not increase it.[10]

Britain's Michael Pitman affirmed:

> Neither observation nor controlled experiment has shown natural selection manipulating mutations so as to produce a new gene, hormone, enzyme system or organ.[11]

Dr. A. E. Wilder-Smith elucidated in his book *The Natural Sciences Know Nothing of Evolution*:

> If water is poured onto a text written in ink, this text will thus be modified or partly smudged; but never is fundamentally new information added to the text in this manner. The chemistry of *mutations* in the genetic code information has an effect similar to that of water on our text. Mutations modify or destroy already existing genetic information, but they never create new information. They never create, for example, an entirely new biological organ such as an eye or ear. Herein lies an error or Neodarwinism, which teaches that fundamentally new information is created by mutations.[12]

Michael Behe, associate professor of biochemistry at Lehigh University, draws this analogy in his 1996 book *Darwin's Black Box*: Suppose you receive a manufacturer's instructions on how to assemble a bicycle. A mutation is like a printing error changing one of the directions from "take a 1/4 inch nut" to "take a 3/8 inch nut." This is very unlikely to give you a better bicycle. But suppose by chance it did, and 3/8 inch nuts became a new standard for the bike. It stretches credibility to say that a series of such errors could, step by step, convert your bicycle into a motorcycle—yet that's what evolution claims.[13]

By the same token, it is nearly impossible to even transform one sentence into another (i.e., "Little Miss Muffet sat on a tuffet" into "Little Bo-Peep has lost her sheep") by changing one letter at a time, and still have it meaningful at each step. The longer the sentence, the more difficult this task. Nor can one organ or animal become another by changing one gene at a time, and still be functioning at each point.

Oh, but wait. What about that example of bacteria resisting antibiotics? Doesn't that prove the mutational theory at least possible? Actually, some bacteria possess a natural genetic capacity to resist certain antibiotics; mutations are not involved. In other cases, mutations cause a structural defect in ribosomes—the cellular constituents that antibiotics like streptomycin attach to. Since the antibiotic doesn't connect with the misshapen ribosome, the bacterium is resistant. But is that evolution? As Spetner notes:

> We see then that the mutation reduces the specificity of the ribosome protein, and that means losing genetic information. . . . Rather than say the bacterium gained resistance to the antibiotic, we would be more correct to say that it lost its sensitivity to it. It lost information. The NDT [neo-Darwinian theory] is supposed to explain how the information of life has been built up by evolution. . . . Information cannot be built up by mutations that lose it. A business can't make money by losing it a little at a time.[14]

In other cases, some mutant bacteria, because they have *defective membranes*, don't absorb nutrients well. Fortuitously for them, that inefficiency also prevents their absorbing antibiotics. And so, in this instance also, they survive better than their normal cousins. But the mutation did not make them stronger, or "evolve" to a higher state.

Likewise, if the world's light suddenly disappeared, blind people might have an advantage over others, since they were already accustomed to operating in darkness. But we cannot then interpret blindness as positive, or representing evolutionary advance.

C. P. Martin, writing in *American Scientist*, made a similar point when he compared x-rays' effect on the body to being kicked and beaten:

> It is quite possible that violent knocking about might dislocate a man's shoulder, and that continued knocking about might actually reduce the previous dislocation. . . . no sane person would cite such a case as this to prove that the results of knocking a man about are not injuries; nor would anyone refer to the result as evidence that knocking a man about can produce an improvement over the normal man. For a truly progressive or evolution-apt mutation must result in an improvement over the normal condition. The truth is that there is no clear evidence of the existence of such helpful mutations. In natural populations endless millions of small and great genic differences exist, but there is no evidence that they arose by mutation.[15]

We could hypothesize further examples of chance benefits from destruction, but they always require exceptional circumstances. Evolutionists, however, have made the exception the rule. Their catalogue of known beneficial mutations is so slim that they must again and again trot out drug-resistant bacteria. From this and other rare instances, such as the development of Concord grapes, they extrapolate an entire world created by mutation.

Even if we allow for an occasional "beneficial mutation," a fish would need more than one to develop an arm or leg—it would require an orchestrated multitude. What are the chances of many mutations occurring together? George Gaylord Simpson, professor of vertebrate paleontology at Harvard, commented:

> Simultaneous appearance of several gene mutations in one individual has never been observed, so far as I know, and any theoretical assertion that this is an important factor in evolution can be dismissed. . . . the probability that five simultaneous mutations would occur in any one individual would be about .0000000000000000000001. This means that if the population averaged 100,000,000 individuals with an average length of generation of only one day, such an event could be expected only once in about

> 274,000,000,000 years—a period about one hundred times as long as the age of the earth.[16]

But couldn't a fish undergo a single mutation contributing to a leg, with more accumulating over eons? Actually, many mutations don't get passed on. Not only do they normally make the individual less fit, but a mutated gene is often recessive—that is, offspring receive the trait only if both parents have it. Examples are sickle cell anemia and cystic fibrosis—children don't get them unless both parents transmit the gene. This is why people are counseled to marry someone from outside their family.

So for our fish friend to perpetuate the "beneficial" mutation creating a bit of a leg, he might have to mate with another fish possessing the same chance mutation. Could such lucky coincidences happen again and again?

And the problem grows. For a mutation to become a new standard, it must spread through the whole species. There's an adage, "Two heads are better than one." Suppose that, by mutation, a man was born with two heads. And say he was smarter than everyone else, and thus able to survive better. Even so, for the trait to become universal, the man would first have to pass it to his children. If the mutation was recessive, this would be impossible unless his wife also had two heads. Then their two-headed offspring would have to mate, and somehow be so attractive or intelligent compared to other humans, that all one-headed people were eliminated. This would surely be a long process—but one that must occur every time any mutation becomes the norm for any species.

Changing species is no easy trick. As the last chapter noted, fossils show them remaining unchanged for eons. Why, then, have some creatures apparently never mutated—while, others, if evolution is true, mutated like crazy?

There are mutations—but not *transmutations*. Zoologist Pierre-Paul Grassé, former president of the French Academy of Sciences, observed:

> The opportune appearance of mutations permitting animals and plants to meet their needs seems hard to believe. Yet the Darwinian theory is even more demanding: A single plant, a single animal would require thousands and thousands of lucky, appropriate events. Thus, miracles would become the rule: events with an infinitesimal probability could not fail to occur.[17]

Grassé's comments bear witness that evolutionists, while rejecting miracles such as God creating life, must conjure miracles of their own to make Darwin's theory work. Spetner has calculated the odds of developing a single new species through mutation, by multiplying the rate of mutation times the chances of a mutation being advantageous, times the chances of the mutation spreading throughout a species, times the number of steps required to form a new species. Though he uses rates generous to evolutionary theory, the resulting odds are one in 3.6 x 10^{2738} (that's a 36 with 2,737 zeroes after it).[18] Mathematicians usually consider anything with odds greater than 1 in 10^{50} effectively impossible. To loosely paraphrase Spetner, believing species develop though mutations is like believing you'll become a millionaire by collecting the pennies you find while out strolling.

So far, incidentally, no evolutionist has been spotted basking in x-rays in hopes of "mutating into a higher state."

CHAPTER 4

Logic Storms Darwin's Gates

Few evolutionists would agree, but by my reckoning, there are now two strikes on Darwinism. Whoa! Here comes a Nolan Ryan fastball!

The problem of half organs

In a popular evolutionary explanation, here's how reptiles evolved into birds: They wanted to eat flying insects that were out of reach. So the reptiles began leaping, and flapping their arms to get higher. Over millions of years, their limbs transformed into wings by increments.[1] In another model, the reptiles were tree-dwellers who leaped. Those who glided well survived and eventually developed wings, but those who glided poorly went kerplunk and were wiped out. In these scenarios, the reptiles' scales sprouted feathers over time, and finally, they became birds.

One problem is that, anatomically, reptilian scales and bird feathers are completely dissimilar. Scales are a tough, thin plate. Feathers are soft and delicate; like hair, they arise from small holes in the skin called follicles; they are held together by a network of little hooks invisible to the naked eye—one eagle feather has over 250,000 of them.

In *Vertebrate History: Problems in Evolution*, Barbara J. Stahl noted: "It is not difficult to imagine how feathers, once evolved, assumed additional functions, but how they rose initially, presumably from reptilian scales, defies analysis."[2]

The theory suffers from an illogical premise that pervades Darwinism. According to natural selection, a physical trait is acquired because it enhances survival. Obviously, flight is beneficial, and one can certainly see how flying animals might survive better than those who couldn't, and thus natural selection would preserve them.

The problem is, wings would have no survival value until they reached the point of flight. Birds' wings and feathers are perfectly designed instruments. Those with crippled or clipped wings cannot fly, and are bad candidates for survival. Likewise, the intermediate creature whose limb was half leg, half wing, would fare poorly—it couldn't fly, nor walk well. Natural selection would eliminate it without a second thought.

The same would hold true for the limbs that Darwinism's fish supposedly developed, or for any body part. Until the organ is operative, it offers no advantage, and natural selection has no reason to favor it. As Stephen Jay Gould asked: "Of what possible use are the imperfect incipient stages of useful structures? What good is half a jaw or half a wing?"[3]

Evolutionists, including Gould, generally try to explain organs' development by hypothesizing that, en route to becoming functional, they must have served some other useful purpose, now unknown to us, and thus survived under natural selection. But this is merely a rationalization based on no evidence.

A classic example of an organ that could not have evolved is the human eye, whose superlative design was not really appreciated until the invention of cameras and other optical instruments dependent on the same principles. For sight to occur, light must pass through the pupil, which automatically adjusts, by widening or contracting, to permit a proper amount of light to enter the eye. It then passes through the lens, which focuses the image on the retina, the light-sensitive area in back of the eye. The retina contains more than 120 million photosensitive cells called rods and cones, which translate light into nerve impulses that reach the brain via the optic nerve.

Vision requires that all of these be working. How then did natural selection make them? Did the lens precede the retina? Did the optic nerve come first? By themselves, none of these constitute vision; they possess no inherent survival value that would cause natural selection to prefer them. To accept evolution, we must believe that chance mutations simultaneously developed each, until one day, by sheer coincidence, all were complete and harmoniously arranged, and vision occurred. The situation troubled Darwin himself, who noted:

> To suppose that the eye with all its inimitable contrivances for adjusting the focus to different distances, for admitting different amounts of light, and for the correction of spherical and chromatic aberration, could have been formed by natural selection, seems, I freely confess, absurd in the highest degree.[4]

Most evolutionary texts avoid discussing eye evolution. One that tried—Gavin de Beer's *Atlas of Evolution*—followed Darwin's own attempted explanation by showing a sequence of eyes of different organisms, starting with the most primitive. But as one observer pointed out, "This mere listing of eyes from various animals, which he neglects (or is unable) to show to be related can carry no conviction for the case for evolution. It would be equally stupid to place a candle, a torch and a searchlight side by side and proceed to advance to a genealogical relationship."[5]

What about color vision? Michael Pitman wrote:

> It is found in several bony fishes, reptiles, birds, bees and primates. Among mammals only primates see in color. Dogs, cats, horses and bulls do not. Fish supposedly evolved the necessary retinal cones to give them color vision, but then lost them. "Re-evolved" by certain unrelated birds and reptiles, they were lost by mammals, but by luck "re-surfaced" in primates. An odd story indeed.[6]

Science has proven the eye far more complicated than was known in Darwin's time. Any organ, of course, may be reduced to its molecular structure. In *Darwin's Black Box,* biochemist Michael Behe describes vision's microscopic physiology. The following material is full of technical words, but this underscores the complexity:

> When light first strikes the retina a photon interacts with a molecule called 11-cis-retinal, which rearranges within picoseconds to trans-retinal. (A picosecond is about the time it takes light to travel the breadth of a single human hair.) The change in the shape of the retinal molecule forces a change in the shape of the protein, rhodopsin, to which the retinal is tightly bound. The protein's metamorphosis alters its behavior. Now called metarhodopsin II, the protein sticks to another protein, called transducin. Before bumping into metarhodopsin II, transducin had tightly bound a small molecule called GDP. But when transducin interacts with metarhodopsin II, the GDP falls off, and a molecule called GTP binds to transducin. . . .
>
> GTP-transducin-metarhodopsin II now binds to a protein called phosphodiesterase, located in the inner membrane of the cell. When attached to metarhodopsin II and its entourage, the phosphodiesterase acquires the chemical ability to "cut" a molecule called cGMP. . . .
>
> When the amount of cGMP is reduced because of cleavage by the phosphodiesterase, the ion channel closes, causing the cellular concentration of positively charged ions to be reduced. This causes an imbalance of charge across the cell membrane that, finally, causes a current to be transmitted down the optic nerve to the brain. The result, when interpreted by the brain, is vision.[7]

So mutations had to engineer, simultaneously, not only the gross anatomical structures of the eye, but its elaborate molecular interactions.

Irreducible complexity

As the complexity of anything increases, the probability of chance creating it decreases. The main point of Behe's book is that biochemistry has proven a number of bodily systems to be irreducibly complex. He says that "design [intelligent creation] is evident when a number of separate, interacting components are ordered in such a way as to accomplish a function beyond the individual components."[8] Gradual change, as Darwin proposed, cannot produce such systems because "any precursor to an irreducibly complex system that is missing a part is by definition nonfunctional."[9]

Blood clotting swings into action when we get a cut. Its multi-step process utilizes numerous proteins, many with no other function besides clotting. Each protein depends on an enzyme to activate it. So which evolved first—the protein or enzyme? Not the protein; it cannot function without the enzyme to switch it on. But why would the enzyme have come first?—without the protein, it serves no purpose. The system is irreducibly complex.

If a person lacks just one clotting factor, as in hemophilia (a mutational disorder), he risks severe bleeding. Furthermore, after a clot forms, the proteins which produced it must be *in*activated by other substances—otherwise the rest of the person's blood would start to coagulate. Step-by-step evolution of clotting is inconceivable: in the trial and error stage, organisms would have either bled to death or clotted to death.

Another example Behe gives: the immune system. In infections, it must distinguish the invading bacterial cells from the body's own cells—otherwise the latter will be attacked (which is the case in "autoimmune" diseases). An antibody identifies the bacterium by attaching to it. In a complex biochemical process, a variety of white blood cells—"killer cells" such as lymphocytes and macrophages—are notified of the bacterium's presence. These travel to the site, and, using the identifying antibody, attack the enemy.

Like blood clotting, this system is irreducibly complex. What evolved first? The killer cells? Without the identifying antibody, they wouldn't know where to attack. But why would the identifier develop first, without anything to notify? If the network evolved gradually, disease would kill the individual before it was perfected.

Behe notes the paucity of articles and books on *how* such biochemical entities evolved. He says that "if you search the scientific literature on evolution, and if you focus your search on the question of how molecular machines—the basis of life—developed, you find an eerie and complete silence. The complexity of life's foundation has paralyzed science's attempt to account for it. . . ."[10] In his research, he found only two very short, highly speculative papers attempting to explain the immune system's evolution on a molecular level.[11] Biologist James Shapiro of the University of Chicago agrees:

> There are no detailed Darwinian accounts for the evolution of any fundamental biochemical or cellular system, only a variety of wishful speculations. It is remarkable that Darwinism is accepted as a satisfactory explanation for such a vast subject—evolution—with so little rigorous examination of how well its basic theses work in illuminating specific instances of biological adaptation or diversity.[12]

The human body, as a total system, is also irreducibly complex. It is difficult to change one part without influencing others. Take the liver: it manufactures bile; detoxifies poisons and wastes; regulates storage and use of glucose, proteins, fats and vitamins; synthesizes blood clotting and immune system factors; and processes breakdown products of old blood cells. Or take the kidneys: they remove wastes through urine production; regulate the body's water content and electrolytes (sodium, calcium, etc.); and support the adrenal glands, which secrete hormones such as adrenaline.

Evolution says every organ developed through chance mutations. But structures like the liver or kidneys cannot change without drastically affecting the rest of the body, with which they maintain a delicate balance. We must assume that by lucky happenstance, each time these organs mutated, other body parts also mutated in harmonious cooperation. The noted writer Arthur Koestler commented:

> You cannot have a mutation A occurring alone, preserve it by natural selection, and then wait a few thousand or million years until mutation B joins it, and so on, to C and D. Each mutation occurring alone would be wiped out before it could be combined with the others. They are all interdependent. The doctrine that their coming together was due to a series of blind coincidences is an affront not only to common sense but to the basic principles of scientific explanation.[13]

And what of the human brain? W. H. Yokel wrote in a promotional letter for *Scientific American* in 1979:

> The deep new knowledge about the brain, gathered at an accelerated rate in recent years, shows this organ to be marvelously designed and capacitated beyond the wonders with which it was invested by ignorant imagination.

> Microelectronics can pack about a million circuits in a cubic foot, whereas the brain has been estimated to pack a million million circuits per cubic foot. Computer switches interact with not more than two other switches at a time, whereas a brain cell may be wired to 1,000 other cells on both its input and output sides. . . .[14]

Yokel called the brain "designed." It has about ten billion neurons (nerve cells) with a thousand trillion connections.[15] Each neuron contains around one trillion atoms. The brain can do the work of hundreds of supercomputers. Building a computer requires great intelligence. Who believes even a simple one could arise by chance? Indeed, the brain is more than a computer—"It is a video camera and library, a computer and communications center, all in one."[16] Yokel added:

> Perhaps the most elusive questions surround the brain functions that make us human—the capacities of memory and learning. Transcending what might be called the hardware of the brain, there comes a software capacity that eludes hypothesis. The number that expresses this capacity in digital information bits exceeds the largest number to which any physical meaning can be attached.[17]

How about our thoughts? Did chance evolve them, too? Darwin critic Phillip Johnson asks:

> Are our thoughts "nothing but" the products of chemical reactions in the brain, and did our thinking abilities originate for no reason other than their utility in allowing our DNA to reproduce itself? Even scientific materialists have a hard time believing *that.* For one thing, materialism applied to the mind undermines the validity of all reasoning, including one's own. If our theories are the products of chemical reactions, how can we know whether our theories are true? Perhaps [evolutionist] Richard Dawkins believes in Darwinism only because he has a certain chemical in his brain, and his belief could be changed by somehow inserting a different chemical.[18]

The animal kingdom also illustrates features too complex for evolution to explain. *Time* magazine, in an article critical of creationism, described the remarkable bombardier beetle:

> Its defense system is extraordinarily intricate, a cross between tear gas and a tommy gun. When the beetle senses danger, it internally mixes enzymes contained in one body chamber with concentrated solutions of some rather harmless compounds, hydrogen peroxide and hydroquinones, confined to a second chamber. This generates a noxious spray of caustic benzoquinones, which explode from its body at a boiling 212 F. What is more, the fluid is pumped through twin rear nozzles, which can be

> rotated, like a B-17's gun turret, to hit a hungry ant or frog with bull's-eye accuracy.[19]

Since the chemicals and enzymes are explosive when mixed in a small space, how could chance have evolved this defense system without blowing the beetle to smithereens?

Or how could natural selection produce the monarch butterfly, which transforms from a caterpillar to a butterfly with two compound eyes, each with 6,000 lenses, and a brain that can decipher 72,000 nerve impulses from the eyes?[20]

Extinct creatures make the problem even more perplexing. Cambrian rocks, dated at over 500 million years, contain fossils of many of the oldest invertebrates known. Among them: the trilobite. It was an arthropod (the broad category of joint-legged animal that includes lobsters and spiders).

According to evolution, Cambrian rocks should contain only "primitive" organisms. The trilobite was anything but. It had a segmented body, legs, gills, antennae, and a complex nervous system. Moreover, according to *Science News*, trilobites had "the most sophisticated eye lenses ever produced by nature."[21] Riccardo Levi-Setti of the University of Chicago writes in his book *Trilobites*:

> In fact, this optical doublet is a device so typically associated with human invention that its discovery in trilobites comes as something of a shock. The realization that trilobites developed and used such devices half a billion years ago makes the shock even greater. And a final discovery—that the refracting interface between the two lens elements in a trilobite's eye was designed in accordance with optical constructions worked out by Descartes and Huygens in the mid-seventeenth century—borders on sheer science fiction. . . . The design of the trilobite's eye lens could well qualify for a patent disclosure.[22]

Like other animals, trilobites lack fossil ancestors showing how they evolved.

No comparison

Mr. Darwin's problems aren't over. Taxonomy is the science that classifies plants and animals, grouping them according to characteristics they share. Swedish botanist Carolus Linnaeus (1707–1778) pioneered the field. He assigned organisms by class, order, genus and species, and developed the two-word naming system, i.e., a cat is *Felis cattus*—belonging to the species *cattus* of the genus *Felis*. The system won universal acceptance. Linnaeus, who believed the Biblical account of creation, saw that the animal kingdom's larger divisions—contrary to evolution—were distinctly divided without overlaps.

Two centuries later, Theodosius Dobzhansky, professor of zoology at Columbia University, essentially confirmed that observation:

> If we assemble as many individuals living at a given time as we can, we notice at once that the observed variation does not form any kind of continuous distribution. Instead, a multitude of separate discrete distributions are found. In other words, the living world is not a single array of individuals in which any two variants are connected by unbroken series of intergrades, but an array of more or less distinctly separate arrays, intermediates between which are absent or at least rare.[23]

Austin H. Clark, the eminent zoologist of the Smithsonian Institution, was no creationist, but he declared:

> The complete absence of any intermediate forms between the major groups of animals, which is one of the most striking and most significant phenomena brought out by the study of zoology, has hitherto been overlooked, or at least ignored. . . .
>
> No matter how far back we go in the fossil record of previous animal life upon the earth we find no trace of any animal forms which are intermediate between the various major groups or phyla.
>
> This can only mean one thing. There can be only one interpretation of this entire lack of any intermediates between the major groups of animals—as for instance between the backboned animals or vertebrates, the echinoderms, the mollusks and the arthropods.
>
> If we are willing to accept the facts we must believe that there never were such intermediates, or in other words that these major groups have from the very first borne the same relation to each other that they bear today.[24]

In the African landscape, we see animals showing little resemblance to each other: lions, zebras, elephants, monkeys, giraffes. If Darwinism is true, why did these creatures adapt to the same general environment so differently? If they are descended from a common ancestor, why don't we see species intermediate between them? Where are the giraffants and eleraffes?

The evolutionist, believing that the various animals branched from each other in the distant past, supposes the intermediates all became extinct. But if that is true, where are their fossils? Suppositions become endless. As W. R. Thompson noted in his introduction to a 1956 edition of *The Origin of Species*:

> Taking the taxonomic system as a whole, it appears as an orderly arrangement of clear-cut entities, which are clear-cut because they are sep-

> arated by gaps. . . . The general tendency to eliminate, by means of unverifiable speculations, the limits of the categories nature presents to us, is the inheritance of biology from the *Origin of Species*. To establish the continuity required by the theory, historical arguments are invoked, even though historical evidence is lacking. Thus are engendered those fragile towers of hypotheses based on hypotheses, where fact and fiction intermingle in an inextricable confusion.[25]

Recent scientific advances validate Linnaeus's original view. Evolutionists had expected that since animals have many similar anatomical features, their genes would also prove similar. After all, it's genes from ancestors that determine our physical characteristics. But molecular biologist Michael Denton notes that "homologous [corresponding] structures are specified by quite different genes in different species."[26] There are exceptions, but they are relatively rare. As Gavin de Beer, former director of the British Museum, acknowledged:

> It is now clear that the pride with which it was assumed that the inheritance of homologous structures from a common ancestor explained homology was misplaced; for such inheritance cannot be ascribed to identity of genes.[27]

Evolutionists have long cited resemblances in the biochemistry of men and animals as proof of their common descent. However, some similarity is not surprising, since we all breathe the same air, eat the same food and drink the same water. Breakthroughs in biological research have in fact underscored the differences. Denton, in *Evolution: A Theory in Crisis*, noted:

> The hope that increased biochemical knowledge would bridge the gap was specifically expressed by many authorities in the 1920s and 30s. But, as in so many other fields of biology, the search for continuity, for empirical entities to bridge the divisions of nature, proved futile. Instead of revealing a multitude of transitional forms through which the evolution of a cell might have occurred, molecular biology has served only to emphasize the enormity of the gap. . . .
>
> [N]o living system can be thought of as being primitive or ancestral with respect to any other system, nor is there the slightest empirical hint of an evolutionary sequence among all the incredibly diverse cells on earth. For those who hoped that molecular biology might bridge the gulf between chemistry and biochemistry, the revelation was profoundly disappointing. . . .
>
> [A]s more protein sequences began to accumulate during the 1960s, it became increasingly apparent that the molecules were not going to provide any evidence of sequential arrangements in nature, but were rather going to reaffirm the traditional view that the system of nature conforms funda-

> mentally to a highly ordered hierarchic scheme from which all direct evidence for evolution is emphatically absent.[28]

Evolution says bacteria evolved successively into fish, amphibians, reptiles, mammals and man. Humans have about 1,000 times more DNA than bacteria—yet salamanders, who are amphibians, have 20 times more DNA than humans.[29] Humans have 30 times more DNA than some insects, but less than half that of certain other insects.[30] More significant are the *qualitative* contradictions. Colin Patterson of the British Museum of Natural History has pointed out that when amino acid sequences for alpha hemoglobins are compared, crocodiles and chickens are more similar than crocodiles and vipers, even though the latter two are both reptiles.[31]

Denton's book analyzes various molecular structures, such as that of cytochrome C, a protein involved in creating cellular energy. It is found in organisms ranging from bacteria to man. Based on cytochrome C, a turtle is closer to a bird than to a snake.[32] Denton reports:

> [A]lthough cytochrome C sequences varied among the different terrestrial vertebrates, all of them are equidistant from those of fish. At a molecular level there is no trace of the evolutionary transition from fish ➝ amphibian ➝ reptile ➝ mammal. So amphibia, always traditionally considered intermediate between fish and the other terrestrial vertebrates, are in molecular terms as far from fish as any group of reptiles or mammals![33]

We're not done yet

Yo, Darwin! More problems. In his *Bioscience* article "Extinction Rates Past and Present," Norman Myers stated:

> Today's rate [of extinction] can be estimated through various analytical techniques to be a minimum of 1000, and possibly several thousand species per year. . . . It generally takes tens of thousands of years for a new terrestrial vertebrate or a new plant species to emerge fully, and even species with rapid turnover rates, notably insects, usually require centuries, if not millennia, to generate a new species.[34]

If thousands of species become extinct every year, but it takes eons to evolve just one, how is evolution mathematically possible? Man's impact on the environment has accelerated modern extinction rates, but is this sufficient to resolve the dilemma?

Let's ask an even more fundamental question: Why aren't fish *today* growing little arms and legs, trying to adapt to land? Why aren't reptiles *today* developing feathers? Shouldn't evolution be ongoing? If organs take eons to build, why does every creature today have complete, functioning parts instead of half-formed ones?

Evolution is not visible in the past, via the fossil record. It is not visible in the present, whether we consider an organism as a whole, or on the microscopic planes of biochemistry and molecular biology, where, as we have seen, the theory faces numerous difficulties. In short, evolution is just not visible. Science is supposed to be based on observations.

PLATE 7. Carolus Linnaeus, founder of modern taxonomy and a creationist

Taxonomy (Classification)—
A Simple Review

(1) **KINGDOM**—one of the large divisions of the living world. Included are the animal and vegetable kingdoms, and those of microscopic organisms.

(2) **PHYLUM**—a major division within a kingdom. Arthropods (joint-legged segmented creatures such as lobsters and insects) and mollusks (soft-bodied creatures usually with shells, such as oysters and snails) are examples of phyla. Man belongs to the phylum of chordates, as do all other back-boned creatures.

(3) **CLASS**—divides a phylum into more similar creatures. Mammals, reptiles and birds are classes within the phylum of chordates.

(4) **ORDER**—a subdivision of class. Among mammals, dogs are in the order of carnivores (flesh-eaters); moles are in the order of insectivores (insect-eaters).

(5) **FAMILY**—similar creatures within an order. Lions, tigers, leopards, and domestic cats all belong to the cat family.

(6) **GENUS**—a further subdivision. Lions and tigers belong to the genus *Panthera*, house cats to the genus *Felis*.

(7) **SPECIES**—classification's bottom rung. Lions (*Panthera leo*) and tigers (*Panthera tigris*) are different species within the same genus. According to evolutionists, man belongs to the species *sapiens* of the genus *Homo* of the family of hominids of the order of primates of the class of mammals of the phylum of chordates of the kingdom of animals.

PLATE 8. The trilobite's complexity defies its Cambrian location.

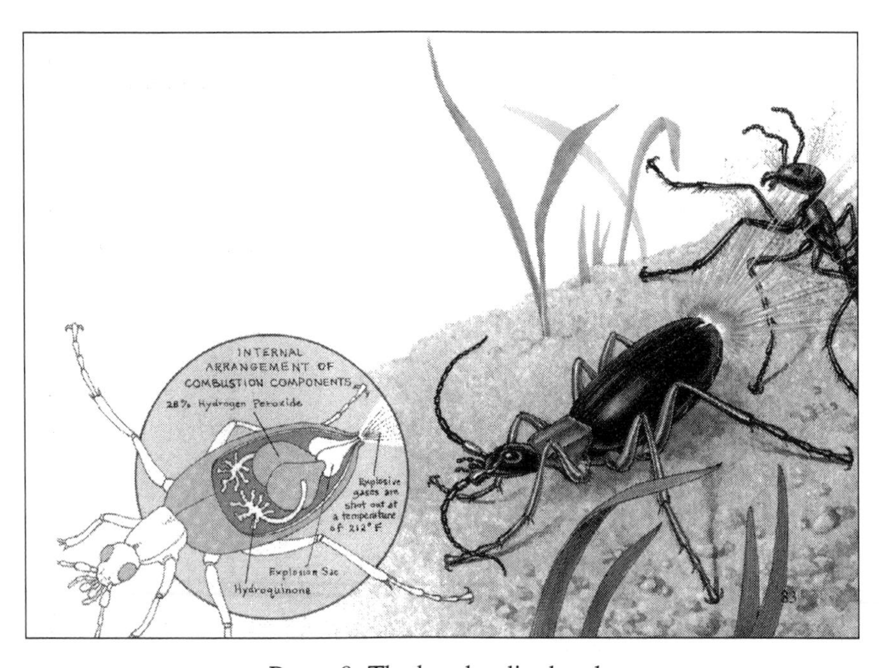

PLATE 9. The bombardier beetle

CHAPTER 5

Origin of the Specious

Are we calling Darwin a bad observer? Are we attempting to treat him with contempt? Are we saying he should be referred to as a pinhead? Certainly not.

Charles Darwin (1809–1882) was the son of a wealthy English physician. As a youngster, he enjoyed collecting shells and bird eggs. He studied medicine for a while at the University of Edinburgh, and later theology at Cambridge, but was generally bored with his education. The event that changed his life came in 1831, when the young man was hired as naturalist on the HMS *Beagle*, an exploring ship on a global excursion that would last five years.

One part of the voyage that greatly intrigued Darwin was a visit to the Galápagos Islands, several hundred miles off the west coast of South America. He noticed that plants and animals differed from island to island. The finches particularly impressed him. Each island seemed to have its own variety with a distinctive beak. Some had large beaks used for cracking seeds; others slender beaks suited to catching insects; some had beaks ideal for probing flowers; one type even had a "wood boring" beak with which it grabbed cactus spines to search inside trees. These finches did not appear to interbreed.

Darwin quite reasonably concluded that the birds had a common ancestor, and that each variety had adapted to its environment. From this and similar observations of other creatures, Darwin inferred that all species—the entire plant and animal kingdoms—had resulted from environmental adaptations over eons. This contradicted the belief, then widely held, that species were immutable. It also contested the Judeo-Christian view that God had created all life forms. It was not until a quarter-century later, in 1859, that Darwin summed up his observations in *The Origin of Species,* but the book caused a sensation and immediately sold out.

Let's give due credit. Darwin was a good student of nature. He correctly observed that animals have variations related to environmental adaptation. His mistake was assuming, without evidence, that an organism has *unlimited* adaptive power, and so could transform into a completely different class of animal—in other words, that a fish could, given enough time and the right environments, become an ape. He wrote of natural selection: "By this process long continued . . . it seems to me almost certain that an ordinary hoofed quadruped might be converted into a giraffe."[1]

Darwin recorded thirteen "species" of finch in the Galápagos. But what did he actually see? Differences in the beaks, tails and plumage of these finches. There is nothing earth-shaking about variation in nature. *People* vary. Like snowflakes, no two are exactly alike (except for identical twins). Some of us have blond hair, some red, black or brown. People differ in many other ways—eye color, skin color, height, etc. None have the same fingerprints. But these distinctions do not separate us into multiple species.

The idea that beak variations constitute species is controversial. Pigeons vary in appearance more than Darwin's finches, yet are considered all part of one species. Dogs differ markedly, but are regarded as one species. Furthermore, although Darwin believed the finches did not interbreed, recent studies in the Galápagos Islands have demonstrated that some of them *do*, resulting in hybrids. Jonathan Weiner, who reported this in *The Beak of the Finch*, noted: "Strange as it seems, these hybrids are the fittest finches on the island."[2] What Darwin observed was not separate species, but ordinary variation *within* a species, no more unusual than the variable dog breeds or different races of man.

Part of the problem: the hazy definition of "species." Darwin himself considered it a loose term and balked at defining it. He said: "I look at the term species as one arbitrarily given for the sake of convenience to a set of individuals closely resembling each other, and that it does not essentially differ from the term variety, which is given to less distinct and more fluctuating forms."[3]

At the famed Scopes trial, Dr. Maynard Metcalf, a zoologist from Johns Hopkins, stated: "In all this discussion I have not used the word 'species.' There are no such things as species in nature. . . . The word species is indefinable, and is used by biologists merely as a convenience, and it has wholly different meanings when applied to different groups of animals and plants."[4]

Also at the trial, Horatio Hackett Newman, a zoologist from the University of Chicago, said: "The species is the unit of classification, but there is serious doubt as to whether species have any reality outside of the minds of taxonomists."[5]

To this day, taxonomists are constantly adjusting the classification of species—the science is inexact. Around 1940, evolutionists Theodosius Dobzhansky and Ernst Mayr expounded a reproductive definition—animals that did not interbreed, they said, were separate species. But this doesn't hold up consistently either. Dogs are considered a different species from coyotes, wolves and jackals—yet occasionally interbreed with any of them. Apparently the definition of species is not "immutable" itself.

As Niles Eldredge of the American Museum of Natural History wrote in *New Scientist*: "Darwin, it has now become commonplace to acknowledge, never really addressed the 'origin of species' in his book of that title."[6] It discussed survival of the fittest—but not *arrival* of the fittest. H. S. Lipson, professor of physics at the University of Manchester, Britain, wrote:

> Darwin's book—*On the Origin of Species*—I find quite unsatisfactory: it says nothing about the origin of *species*; it is written very tentatively, with a special chapter on "Difficulties on theory"; and it includes a great deal of discussion on why evidence for natural selection does *not* exist in the fossil record. . . .[7]

Zoologist Harold G. Coffin concluded that Darwin related a partial truth:

> We must admit that Darwin *did* see different variations from one island to another in the Galápagos. And he *did* see evidences that made it necessary for him to discard a belief that living things did not change. But they were relatively minor and he had no compelling evidence that forced him to believe in limitless transformation. Darwin made a common mistake—that of "either-or." *Either* species were fixed *or* unlimited change occurred. But the truth lies between the two extremes.[8]

Marjorie Grene, the well-known historian of science at the University of California, Davis, observed:

> There are, indeed, all the minute specialized divergences like those of the Galápagos finches which so fascinated Darwin; it is their story that is told in the *Origin* and elaborated by the selectionists today. But these are dead ends, last minutiae of development; it is not from them that the great massive novelties of evolution could have sprung. . . . And if one returns to read *Origin* with these criticisms in mind, one finds, indeed, that for all the brilliance of its hypotheses, for all the splendid simplicity of the "mechanism" by which it "explains" so many and so varied phenomena, it simply is not about the origin of species, let alone of the great orders and classes and phyla, at all. Its argument moves in a different direction altogether, in the direction of minute, specialized adaptations, which lead, unless to extinction, nowhere.[9]

Has anyone ever witnessed the creation of a new species? Geneticist Thomas Hunt Morgan, who won a Nobel Prize for his work on heredity, acknowledged: "Within the period of human history, we do not know of a single instance of the transformation of one species into another, if we apply the most rigid and extreme tests used to distinguish wild species from each other."[10]

Colin Patterson, director of the British Museum of Natural History, said: "No one has ever produced a species by mechanisms of natural selection. No one has gotten near it and most of the current argument in neo-Darwinism is about this question."[11]

Darwin himself could not document a bona fide case. As we have noted, he wrote: "We cannot prove that a single species has changed."[12]

We won't push this point too hard because, again, if "species" is defined very flexibly, one could argue that new ones originate. But this would simply mean variation, as when dogs are bred into different types—evolution in a horizontal, not vertical, sense.

Given our context, I would be remiss if I did not mention a case frequently cited by evolutionists as living proof of Darwin's theory—the peppered moth. The Empire Strikes Back! Harvard profs in Darth Vader helmets!

The peppered moth is a British species that comes in white and black varieties. Before 1850, the black ones were quite rare—nearly all peppered moths were white. The industrial revolution changed this. Peppered moths rest on tree trunks during the day. Pollution killed the lichen on trees and blackened them with soot. As a result, preying birds could better see the white moths. The black moths, however, previously quite visible, were now camouflaged. By 1895, in a role reversal, 98 percent of peppered moths were black. Recent pollution controls, however, have caused a resurgence of the light-colored ones.

Darwinists hail this as evolution occurring before our eyes. Certainly, it is a valid example of natural selection, and the fittest surviving. But does it constitute evolution? The white and black moths are varieties of the same species. Both existed before pollution brought a shift. No new species formed. Now if they had turned into Godzilla's old nemesis Mothra, we'd be more impressed. I again quote the eminent French zoologist Grassé:

> The "evolution in action" of J. Huxley and other biologists is simply the observation of demographic facts, local fluctuations of genotypes, geographical distributions. Often the species concerned have remained practically unchanged for hundreds of centuries! Fluctuations as a result of circumstances, with prior modification of the genome, does not imply

> evolution, and we have tangible proof of this in many panchronic [unchanging] species. . . .[13]

Breeders of dogs and horses have isolated animals with traits they thought desirable. This has given rise to new varieties, but not species. In nature's parallel, a few animals sometimes become geographically isolated. The population they sire inherits their particular genes and thus their characteristics. That is essentially what happened to Darwin's finches—after becoming sequestered on islands, they individuated from other finches, but no "new species" emerged.

Darwin, however, as well as modern evolutionists, have cited breeding experiments as evidence for evolution. Darwin expected that breeders would, in time, develop genuine new species. They haven't. Francis Hitching summed the situation well:

> Every series of breeding experiments that has ever taken place has established a finite limit to breeding possibilities. Genes are a strong influence for conservatism, and allow only modest change. Left to their own devices, artificially bred species usually die out (because they are sterile or less robust) or quickly revert to the norm.[14]

No one has ever grown black tulips or blue roses, because the genes are not there. Luther Burbank, the famed American plant breeder, said:

> I know from my experience that I can develop a plum half an inch long or one 2½ inches long, with every possible length in between, but I am willing to admit that it is hopeless to try to get a plum the size of a small pea, or one as big as a grapefruit. . . . I have roses that bloom pretty steadily for six months in the year, but I have none that will bloom twelve, and I will not have. In short, there are limits to the development possible, and these limits follow a law. . . . plants and animals all tend to revert, in successive generations, toward a given mean or average.[15]

The more a species is pushed from the norm, the harder it gets. Evolutionist Ernst Mayr demonstrated this when he tried to modify fruit flies through breeding. The flies normally averaged 36 bristles on their bodies. When the insects were bred down to 25 bristles, they became sterile. Mayr also bred the flies upwards—but when they reached 56 bristles, they again became sterile.[16] If controlled conditions and intelligent planning cannot transform a species, how much less could natural selection, operating under chance conditions without intelligence?

Suppose we wanted elephants with long trunks. So we paired off long-trunked elephants, and allowed only their offspring with the longest trunks to mate, and so forth. Well, we would get some long trunks that way, but never trunks a thousand feet long. Genes impose limits. But Darwin, ignorant of genetics, declared:

> I can see no limit to the amount of change, to the beauty and complexity of the co-adaptations between all organic beings, one with another and with their physical conditions of life, which may have been effected in the long course of time through nature's power of selection, that is by the survival of the fittest.[17]

Modern Darwinists, of course, have theorized that genes will transcend their limits through mutation. But as Chapter Three illustrated, this argument is unsupportable. Evolutionist Richard Goldschmidt, one of the twentieth century's most eminent geneticists, acknowledged:

> It is true that nobody thus far has produced a new species or genus, etc., by macromutation. It is equally true that nobody has ever produced even a species by the selection of micromutations.[18]

Plant and animal breeders don't manufacture new genes; they simply use genetic information that is already there. In fact, their practices result in *lost* information. When you isolate, say, bulldogs, allowing them to mate only with other bulldogs, they stay bulldogs. They have lost the capacity to look any other way. Thus the creation of special varieties, such as Darwin's finches, means genetic information is gone—not gained, as evolution claims.

We might borrow Goldschmidt's use of "macro" and "micro" to note that, while *micro*evolution does occur—meaning minor adaptations and variations within a species—there is no such thing as *macro*evolution, or conversion of one animal type into another, into a new genus or other broad classification level. Goldschmidt would have known—he bred gypsy moths for twenty years and a million generations.[19] All he ever got was more gypsy moths.

PLATE 10. The HMS *Beagle*

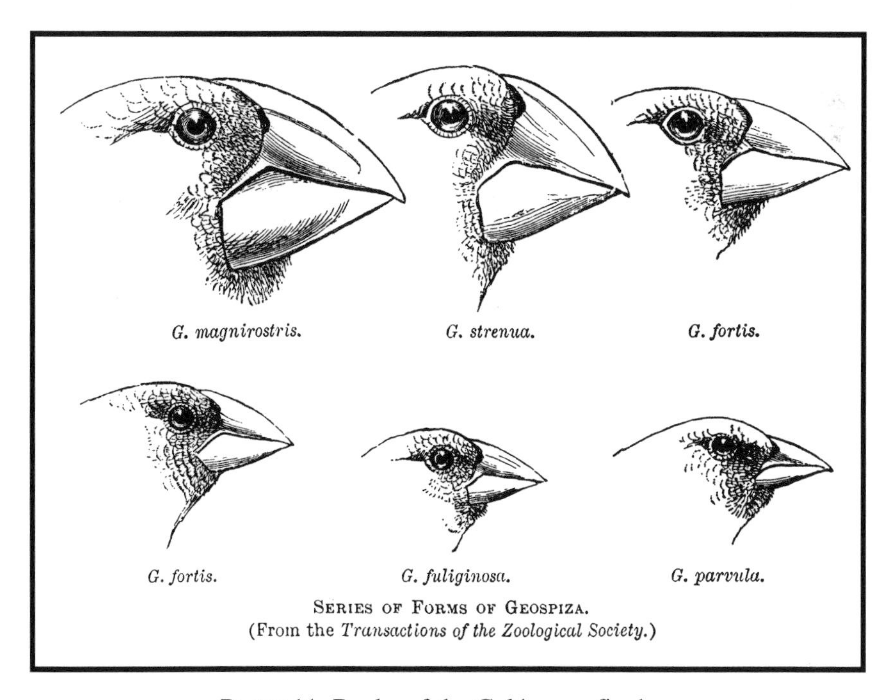

PLATE 11. Beaks of the Galápagos finches

PLATE 12. Charles Darwin in 1854

CHAPTER 6

Darwin vs. Design

Biologist Gary Parker has written:

> Imagine that you are walking along a creek on a lazy summer afternoon, idly kicking at the pebbles along the bank. Occasionally you reach down to pick up a pebble that has an unusual shape. One pebble reminds you of a cowboy boot.
>
> As you roll the pebble around in your hand, you notice that the softer parts are more worn away than the harder parts. . . . Despite some appearance of design, the boot shape of the tumbled pebble is clearly the result of time, chance, and the processes of weathering and erosion.
>
> But then your eye spots an arrowhead lying among the pebbles. Immediately it stands out as different. In the arrowhead, chip marks cut through the hard and soft parts of the rock equally. . . . In the arrowhead, we see matter shaped and molded according to a design that gives the rocky material a purpose.[1]

Now someone reading this book is probably thinking, "OK, Jack, I get it. You're going to lay some 'God must have designed the universe' trip on us. Well, I'm not buying it. Even an arrowhead, given enough time, could have been formed by chance."

Let's up the ante. Say we're walking along in South Dakota, and we notice the faces carved on Mount Rushmore. Will anyone give me odds on *that* being produced by erosion? I think most people would agree that, even in a billion years, it wouldn't happen.

The biology of human beings is far more complex than the Mount Rushmore sculpture. How then could chance create *them* in a billion years?

British philosopher and theologian William Paley (1743–1805) penned the most famous "walking along" argument. He reasoned that if his foot struck a rock during a stroll, well, perhaps the rock had always lain there. But what if he stumbled upon a watch? Could it have arisen by chance? Looking at the timepiece, with its intricate mechanisms, it would be clear that someone had made it. And if the watch had a designer, what about the body's infinitely more complex organs? Paley's "watchmaker" argument held that *design proves a designer*. When we see handwriting on a wall, or a dress in a shop, we know they resulted from intelligence, not chance.

The argument was known well before Paley's time. The Roman philosopher Cicero said:

> When you look at a picture or a statue, you recognize that it is a work of art. When you follow from afar the course of a ship, you do not question that its movement is guided by a skilled intelligence. When you see a sundial or a water-clock, you see that it tells the time by design and not by chance. How then can you imagine that the universe as a whole is devoid of purpose and intelligence when it embraces everything, including these artifacts themselves and their artificers?[2]

One complaint posed by atheists (I know, I used to be one) runs like this: "It's too easy. Ask a person who believes in the Bible why various things exist, he can't answer your questions, so he just stands there with his mouth hanging open, repeating, 'Oh, well, duh, God must have created it that way!'"

Of course, a redundant reply is somewhat obligatory if God really created the universe. And evolutionists have their own pat answer: natural selection, which is so plastic that it can be made to explain anything.

For example, a reptile's lower jaw has several bones, while the mammalian jaw consists of a single bone. Reptiles have one bone in each ear, mammals three. If reptiles became mammals, how did this transformation take place? Evolutionists claim that, over time, bones from the reptilian jaw must have migrated to the ear and become mammalian ear bones. Yet there are no fossils of numerical intermediates (i.e., animals with two ear bones, or two or three bones in the lower jaw). And how did reptiles hear and chew during the transition? This is what I mean when I call natural selection plastic. Nevertheless, such theories tend to become "fact" once enunciated.

Darwin did the same thing in saying our ape-like ancestors eventually lost their body hair because they preferred mates with less hair. (Do we still have smelly armpit hair because they thought that was a turn-on?) There is no proof. Anyone can invent a hypothesis to explain something—that's different from demonstrating it to be a fact, and certainly no more substantive than saying that God created things that way.

Did giraffes really develop long necks because they lived around high vegetation, causing the extinction of shorter-necked giraffes? How then did young giraffes survive? Isn't it more likely that, facing such an environment, giraffes would have simply migrated to where food was more accessible? Colin Patterson of the British Museum of Natural History noted:

> It is easy enough to make up stories of how one form gave rise to another, and to find reasons why the stages should be favoured by natural selection. But such stories are not part of science, for there is no way of putting them to the test.[3]

Gould et al. wrote in *Paleobiology*:

> Paleontologists (and evolutionary biologists in general) are famous for their facility in devising plausible stories; but they often forget that plausible stories need not be true.[4]

And I again quote France's Pierre-Paul Grassé:

> Today, our duty is to destroy the myth of evolution, considered as a simple, understood, and explained phenomenon which keeps rapidly unfolding before us. Biologists must be encouraged to think about the weaknesses of the interpretations and extrapolations that theoreticians put forward or lay down as established truths. The deceit is sometimes unconscious, but not always, since some people, owing to their sectarianism, purposely overlook reality and refuse to acknowledge the inadequacies and the falsity of their beliefs.[5]

While evolutionists can think up logical-sounding reasons for why natural selection produced certain things, many phenomena resist such rationalization. Canadian biologist Ludwig Bertalanffy told a symposium:

> I, for one, in spite of all the benefits drawn from genetics and the mathematical theory of selection, am still at a loss to understand why it is of selective advantage for the eels of Comacchio to travel perilously to the Sargasso sea, or why *Ascaris* has to migrate all around the host's body instead of comfortably settling in the intestine where it belongs; or what was the survival value of a multiple stomach for a cow when a horse, also vegetarian and of comparable size, does very well with a simple stomach; or why certain insects had to develop those admirable mimicries and protective colorations when the common cabbage butterfly is far more abundant with its conspicuous white wings. One cannot reject these and innumerable similar questions as incompetent; if the selectionist explanation works quite well in some cases, a selectionist explanation cannot be refused in others.
>
> In current theory, a speculative "may have been" or "must have been" (expressions occurring innumerable times in selectionist literature) is accepted in lieu of an explanation which cannot be provided. . . . in my opinion, there is no scintilla of scientific proof that evolution in the sense of progression from less to more complicated organisms had anything to

> do with better adaptation, selective advantage or production of larger offspring.[6]

Noticing that children resemble their parents, Charles Darwin correctly observed that characteristics are inherited. But he inflated the conclusion. He asked: "What can be more curious than that the hand of a man, formed for grasping, and that of a mole for digging, the leg of a horse, the paddle of the porpoise, and the wing of a bat should all be constructed on the same pattern, and should include similar bones, in the same relative positions?" He summed these and other observations by quoting the zoologist W. H. Flower: "Is it not powerfully suggestive of true relationship, of inheritance from a common ancestor?"[7] Since the animal that most resembles man is the ape, he decided they had a joint line of descent.

Undeniably, various creatures, including man, have common physical structures: two eyes, four limbs, a heart, brain, and so forth. But do these things exist collectively because of ancestry—or because the design is efficient?

A Toyota Camry strongly resembles a Nissan Sentra. Both have a transmission, headlights, brakes, four wheels, and countless other corresponding features. Does this imply a genealogical relationship? They share these components because the design is efficient. Just try driving a car with two wheels.

Like animals, cars are not exactly alike. They vary in size, shape, color, style, and accessories. But there are *basic principles of design* shared by all vehicles, ranging from a child's remote-control racer all the way up to a tractor-trailer, just as there are in animals from a mouse to an elephant. Mark Ridley, a research fellow at Oxford University, observed in *The Problems of Evolution*:

> Any set of objects, whether or not they originated in an evolutionary process, can be classified hierarchically. Chairs, for instance, are independently created; they are not generated by an evolutionary process: but any given list of chairs could be classified hierarchically, perhaps by dividing them first according to whether or not they were made of wood, and then according to their color, by date of manufacture, and so on. The fact that life can be classified hierarchically is not, in itself, an argument for evolution.[8]

Darwin believed that bone similarities prove a mutual line of descent. Let us then observe that the human hand and foot strongly resemble each other. Each has five digits. The thumb and large toe each have two bones, while the smaller digits all have three. How do we explain these uncanny parallels between hands and feet? A common origin? Not according to Darwinism's fish-to-land theory. As Michael Denton notes: "There is no doubt that in terms of evolution the fore- and hindlimbs must have risen independently, the former supposedly evolving from the pectoral fins of a fish, the latter from the pelvic fins." And he adds that "no evolutionist claims that the hindlimb evolved from the forelimb, or that the hindlimbs and forelimbs evolved from a common source."[9] Darwinism can only explain the resemblance as coincidence.

And why do so many things come paired?—eyes, ears, nostrils, arms, legs, hands, feet, breasts, lungs, kidneys, gonads, lobes of the brain, wings on birds. Natural selection's answer: "Why, animals with two eyes survived better than those with one." But if it is just a survival matter, why stop at two? Why not three eyes, or eight?

These organs are not only paired, but are mirror images of each other. Symmetry is aesthetic; artists use it to convey beauty. It speaks for design more than chance and mutation.

If we can acknowledge design in an arrowhead—as an anthropologist surely will—why not in a rose's lovely petals and fragrance, or a peacock's feathers? The evolutionist responds, "Oh, beauty makes animals more attractive to mates—it has survival value for natural selection." But do animals see beauty as we do? And what of pretty tropical fish that swim at depths where their colors cannot even be seen?[10]

Despite all his intelligence, man has trouble duplicating what "chance and mutation" have supposedly evolved. A few decades ago, some doctors urged mothers to stop breast feeding, claiming baby formulas would be superior. We all know that has proved false. When an infant is born prematurely, the mother's milk even changes to meet its special needs. But under the "survival of the fittest" law, nature should supposedly weed out and eliminate the premature infant, since it is weaker than others.

Darwin pictured a tooth-and-claw world, where creatures compete in the "struggle for existence," and only the strong survive. Evolutionists are fond of reciting offensive and defensive mechanisms that natural selection has endowed certain animals with—the tiger's claws, the moose's antlers, the bee's stinger.

But much doesn't fit the picture. How about a simple banana, which, according to Darwinism, probably arose from the same one-celled ancestor as the tiger? If its sole purpose is survival, why did it develop no weaponry, such as barbs, poisonous fumes, or a noxious taste, to prevent its being eaten? Of course, fruits often spread seeds through being eaten, which perpetuates the species, but dare we not say the purpose of a banana, cantaloupe, or any other fruit, *is* to be eaten, just as the rose is there for us to behold and smell?

Much of nature operates in harmony and cooperation. Bees help flowers pollinate. Robert MacArthur, in an extensive study of various types of warblers, observed that they live in peaceful coexistence, without invading each other's territories, even though they eat the same food.[11] In the Southwestern U.S., the yucca plant depends on the yucca moth for fertilization, and the moth's larva depend on the plant for food[12]—even though insects supposedly evolved millions of years before flowering plants. F. W. Went noted in *Scientific American*:

> In the desert, where want and hunger for water are the normal burden of all plants, we find no fierce competition for existence, with the strong crowding out the weak. On the contrary, the available possessions—space, light, water and food—are shared and shared alike by all. If there is not enough for all to grow tall and strong, then all remain smaller. This factual picture is very different from the time-honored notion that nature's way is cut-throat competition among individuals.[13]

Dr. Gary Parker describes cooperation in the ocean:

> Nowadays, there are many large fish with sharp teeth that roam the oceans seeking what they may devour. But as they feed on smaller fish and shrimp, their mouths begin to accumulate food debris and parasites. Lacking recourse to a toothbrush, how is such a fish going to clean its teeth?
>
> For several kinds of fish, the answer is a visit to the local cleaning station. These are special areas usually marked by the presence of certain shrimp and small, brightly colored fish. . . .
>
> Often fresh from chasing and eating other small fish and shrimp, a predatory fish may periodically swim over to take his place in line (literally!) at the nearest cleaning station. When his turn comes, he opens his mouth wide, baring the vicious-looking teeth.
>
> You might suspect, of course, that such a sight would frighten off the little cleaner fish and shrimp. But, no. Into the jaws of death swim the little cleaners. Now even a friendly dog will snap at you if you try to pick off a tick, and it probably irritates the big fish to have a shrimp crawling around on its tongue and little fish picking off parasites in the soft tissues of the mouth. But the big fish just hovers there, allowing the cleaners to do their work. It even holds its gill chamber open so that the shrimp can crawl around on the gill filaments picking off parasites!
>
> At the end of all this cleaning, the second miracle occurs. You might think the fish might respond, "Ah, clean teeth; SNAP, free meal!" But, no. When the cleaning is done, the big fish lets the little cleaner fish and shrimp back out again. Then the big fish swims off—and begins hunting again for little fish and shrimp to eat![14]

If only the strong survive, why does the skittish and vulnerable chicken live, but not the *T. rex*? What about sheep, many of which, if confronted by predators such as wolves, do not run or fight, but freeze and allow themselves to be attacked?

Darwinists may object that "survival of the fittest" does not necessarily imply killer attributes—only the ability to survive in one's niche. But this becomes a *tautology*—a circular argument. One asks: "Who survives?" The answer: "The fittest." So one asks: "Who is the fittest?" The answer: "Those who survive." It explains little. Philosophy professor Gregory Alan Pesely noted:

> One of the most frequent objections against the theory of natural selection is that it is a sophisticated tautology. . . .
>
> What is most unsettling is that some evolutionary biologists have no qualms about proposing tautologies as explanations. One would immediately reject any lexicographer who tried to define a word by the same word, or a thinker who merely restated his proposition, or any other instance of gross redundancy; yet no one seems scandalized that men of science should be satisfied with a major principle which is no more than a tautology.[15]

Fitness does not always mean survival. The smartest, most resourceful persons are *not* necessarily those who leave the most offspring. So in recent years, evolutionists reduced the definition of "fitness" to simply "those who leave the most offspring." But even this entails a rather circular argument. As geneticist Conrad Waddington of Edinburgh University noted:

> There, you do come to what is, in effect, a vacuous statement: Natural selection is that some things leave more offspring than others; and you ask, which leave more offspring than others; and it is those that leave more offspring; and there is nothing more to it than that.[16]

(One wag asked: If the fittest creatures are those who leave the most offspring, why aren't rabbits now rulers of the Earth?)

I have heard this complaint: "The Biblical account of creation is not a competent alternative to natural selection, because it teaches that species are fixed and immutable." This is a widespread falsehood. The Bible only says, in the first book of Genesis, that plants and animals reproduce "after their kinds." But what did "kinds" mean? Was God using Dobzhansky and Mayr's definition of species, or rather, something anyone can understand? Cats are always cats; dogs are always dogs. The Bible doesn't say a finch's beak can't vary in shape, any more than it says people's noses can't vary.

Another common objection to the design argument is: "Even if God did create the world, none of us were around then to see it happen. Since science deals only with observable events, Biblical creation must be excluded from scientific consideration."

But un momento. None of us were around to see the Big Bang, either. And as Chapter Two discussed, the best evidence for ancient life's history—the fossil record—offers little if any support for Darwinism. Evolution, meaning not just beak variations, but an animal's transformation into a completely different type, cannot be observed, either in the present or from the past. Thus evolution depends on faith as much as creation does. L. Harrison Matthews, long director of the London Zoological Society, noted in 1971:

> Belief in the theory of evolution is thus exactly parallel to belief in special creation—both are concepts which believers know to be true but neither, up to the present, has been capable of proof.[17]

So it is ironic that many modern evolutionists demand that Darwinism be accepted as fact. Darwin himself considered it only a theory, not a law, and confided his doubts in a letter to a colleague in 1858, the year before *Origin*'s publication:

> Thank you heartily for what you say about my book; but you will be greatly disappointed; it will be grievously too hypothetical. It will very likely be of no other service than collating some facts; though I myself think I see my way approximately on the origin of the species. But, alas, how frequent, how almost universal it is in an author to persuade himself of the truth of his own dogmas.[18]

The following excepts, from a letter to Darwin by his wife Emma, about his work and religious views, are also revelatory:

> Your mind and time are full of the most interesting subjects and thoughts of the most absorbing kind, (viz. following up your own discoveries) but which make it very difficult for you to avoid casting out as interruptions other sorts of thoughts which have no relation to what you are pursuing, or to be able to give your whole attention to both sides of the question. . . .
>
> It seems to me that the line of your pursuits may have led you to view chiefly the difficulties on one side, and that you have not had time to consider and study the chain of difficulties on the other; but I believe you do not consider your opinion as formed.[19]

Many scientists are recognizing that Darwinism is not only suppositional, but false. In *Science Digest*, Zygmunt Litynski summarized the conclusions of Aimé Michel, who had interviewed several French biologists:

> The classical theory of evolution in its strict sense belongs to the past. Even if they do not publicly take a definite stand, almost all French spe-

cialists hold today strong mental reservations as to the validity of natural selection.[20]

Arthur Koestler noted in 1978:

> In the meantime, the educated public continues to believe that Darwin has provided all the relevant answers by the magic formula of random mutation plus natural selection—quite unaware of the fact that random mutations turned out to be irrelevant and natural selection a tautology.[21]

Norman Macbeth wrote in *American Biology Teacher*:

> Darwinism has failed in practice. The whole aim and purpose of Darwinism is to show how modern forms descended from ancient forms, that is, to construct reliable phylogenies (genealogies or family trees). In this it has utterly failed. . . . Darwinism is not science.[22]

Swedish biologist Søren Løvtrup declared in his book *Darwinism: The Refutation of a Myth*:

> I suppose that nobody will deny that it is a great misfortune if an entire branch of science becomes addicted to a false theory. But this is what has happened in biology; for a long time now people discuss evolutionary problems in a peculiar "Darwinian" vocabulary—"adaptation," "selection pressure," "natural selection," etc.—thereby believing that they contribute to the *explanation* of natural events. They do not, and the sooner this is discovered, the sooner we shall be able to make real progress in our understanding of evolution.[23]

As natural selection's significance crumbles, the possibility of God, creation and design is again making a wedge in scientific circles. In a 1998 cover story entitled "Science Finds God," *Newsweek* noted:

> The achievements of modern science seem to contradict religion and undermine faith. But for a growing number of scientists, the same discoveries offer support for spirituality and hints of the very nature of God. . . . According to a study released last year, 40 percent of American scientists believe in a personal God—not merely an ineffable power and presence in the world, but a deity to whom they can pray.[24]

Though not finding Christianity persuasive, Paul Davies wrote in *New Scientist*:

> The temptation to believe that the Universe is the product of some sort of *design*, a manifestation of subtle aesthetic and mathematical judgement, is overwhelming. The belief that there is "something behind it all" is one that I personally share with, I suspect, a majority of physicists.[25]

John Polkinghorne, who left a career as a distinguished physicist at Cambridge University to become an Anglican priest, observes:

> When you realize that the laws of nature must be incredibly finely tuned to produce the universe we see, that conspires to plant the idea that the universe did not just happen, but that there must be a purpose behind it.[26]

Charles Townes, who shared the 1964 Nobel Prize in Physics for discovering the principles of the laser, says:

> As a religious person, I strongly sense . . . the presence and actions of a creative being far beyond myself and yet always personal and close by.[27]

Such comments would have seemed mild to an older scientist like Max Planck, Nobel Prize winner and founder of modern physics, who wrote:

> There is evidence of an intelligent order of the universe to which both man and nature are subservient. . . . Wherever we look, we find no evidence as far as we can see of any conflict between science and religion, but only complete agreement on the decisive issues instead. . . . Side by side, science and religion wage a constant, continuing and unrelenting struggle against skepticism and dogmatism, against disbelief and superstition. The battle cry, the goal of this struggle, has been and will always be: Forward to God.[28]

Author David Raphael Klein may have said it best:

> Anyone who can contemplate the eye of a housefly, the mechanics of human finger movement, the camouflage of a moth, or the building of every kind of matter from variations in arrangement of proton and electron, and then maintain that all this design happened without a designer, happened by sheer, blind accident—such a person believes in a miracle far more astounding than any in the Bible.[29]

CHAPTER 7

Vegas Odds on Life

At the time *The Origin of Species* was published, about half of all surgical patients typically died from infections, which physicians assumed arose spontaneously. It was the work of Louis Pasteur (1822–1895) and Joseph Lister (1827–1912) that proved infections are always spread, leading to the modern concept of sterile technique.

Darwinists were slow to accept this discovery, however, because evolution, ruling out creation by God, requires that life can arise by itself. Germany's Ernst Haeckel, one of the great nineteenth-century popularizers of Darwinism, continued to push the old concept of "spontaneous generation." He predicted that spontaneously generated bacteria would be discovered. In 1868, he even published speculative drawings of such. Within a year, British zoologist Thomas Huxley, whose strong advocacy of evolution earned him the nickname "Darwin's bulldog," claimed to have discovered Haeckel's bacteria in mud samples taken from the ocean's depths. Huxley dubbed his find *Bathybius haeckelii* in Haeckel's honor, and it was widely reported that spontaneous generation had been proven.[1]

Microscopic examination by a chemist, however, ascertained that Huxley's "bacteria" were just a chemical precipitate. All other efforts to demonstrate spontaneous generation of life have failed.

Natural selection attempts to explain how living things change. But it doesn't explain how they *began*, a gaping hole in evolutionary theory. Today it is commonly taught that life arose from nonliving materials in a "primordial soup." Charles Darwin mused:

> It is often said that all the conditions for the first production of a living organism are now present which could ever have been present. But if (and oh! what a big if!) we could conceive in some warm little pond, with all sorts of ammonia and phosphoric salts, light, heat, electricity, etc., present, that a protein compound was chemically formed ready to undergo still more complex changes, at the present day such matter would be instantly

> devoured or absorbed, which would not have been the case before living creatures were formed.[2]

In Darwin's day, cellular complexity was unappreciated. Haeckel called the cell a "simple little lump of albuminous combination of carbon."[3] It was assumed that chemicals could combine to form one with relative ease. However, as scientific advances showed cells were anything but simple, the odds against their chance formation increased. Even more deadly to the theory: oxygen, present everywhere, would destroy amino acids (the building blocks of proteins, which are the main components of cells).

In the twentieth century, Soviet biochemist A. I. Oparin further developed the evolutionary theory of origins. He proposed that there was no free oxygen in the ancient Earth's atmosphere. Other gases, he said, such as hydrogen, methane and ammonia might have existed, but not oxygen. Combined with ever-increasing estimates for the Earth's age, the probability for chance formation of life was boosted. Oparin summarized his ideas in his 1936 book *The Origin of Life*.

Oparin's theory was ostensibly validated by the famous experiments of Stanley Miller in 1953. Miller, a University of Chicago graduate student, created an apparatus in which he combined water with hydrogen, methane, and ammonia (the gases Oparin had discussed) and subjected the mixture to electric sparks. After a week, he discovered that some amino acids had formed in a trap in the system. It was conjectured that in the primitive Earth, lightning (corresponding to Miller's electricity) could have struck a similar array of chemicals and produced amino acids.

Experiments such as Miller's have created headlines proclaiming that "life has been created" in laboratories. Let us be clear: although scientists have synthesized some organic chemicals, no one has ever synthesized life.

The Oparin-Miller model is probably evolution's most widely accepted theory of origins, but it faces a number of problems:

(1) It is doubtful that the young Earth was oxygen-free. The *Bulletin of the American Meteorological Society* noted in 1982:

> Recent photochemical calculations by atmospheric researchers at Langley [Research Center] were presented at an international scientific conference last fall. They state that, at the time complex organic molecules (the precursors of living systems) were first formed from atmospheric gases, the earth's atmosphere was not composed primarily of methane, ammonia, and hydrogen as was previously supposed. . . . Oxygen in the earth's early atmosphere may have been up to one million times greater than anyone ever thought.[4]

Langley researcher Joel S. Levine stated:

> Many of the figures we used in our computer simulations are conservative. . . . In the case of our calculated oxygen levels, one bit of evidence from the early geological record supports our conclusion. It was puzzling, but geologists know from their analyses of the oldest known rocks that the oxygen level of the early atmosphere had to be much higher than previously calculated. Analyses of these rocks, estimated to be more than 3.5 billion years old, found oxidized iron in amounts that called for atmospheric oxygen to be at least 110 times greater and perhaps up to one billion times greater than otherwise accepted.[5]

That same year, Clemmey and Badham confirmed in *Geology*:

> Recent biological and interplanetary studies seem to favor an early oxidized atmosphere rich in CO_2 and possibly containing free molecular oxygen. The existence of early red beds [sand deposits stained with the rusty color of oxidized iron], sea and groundwater sulphate, oxidized terrestrial and sea-floor weathering crusts, and the distribution of ferric iron in sedimentary rocks are geological observations and inferences compatible with the biological and planetary predictions. It is suggested that from the time of the earliest dated rocks at 3.7 billion years ago, Earth had an oxygenic atmosphere.[6]

(2) But let's assume Oparin and Miller were right, and the early atmosphere *had* no oxygen. Without oxygen, no ozone layer would exist to protect us from the sun's cosmic rays. Carl Sagan pointed out that, subject to that much ultraviolet light, today's organisms would receive a lethal dose of radiation in just 0.3 seconds.[7] How could the fragile beginnings of life have survived in such an environment?

(3) Darwin spoke of life beginning in a "warm little pond." One might visualize ponds having the right chemical makeup—but they wouldn't last for the eons required to develop living organisms. Therefore, modern evolutionists like Oparin speculated that entire oceans constituted the "primordial soup." But that gums up Miller's experiment, because an ocean would need countless tons of amino acids to get the right concentration of chemicals—and would lightning bolts produce that much?

(4) If a vast primordial soup existed, geology should corroborate it. But as Michael Denton notes: "If the traditional story is true, therefore, there must have existed for many millions of years a rich mixture of organic compounds in the ancient oceans and some of this material would likely have been trapped in the sedimentary rocks lain down in the seas of those remote times. Yet rocks of great antiquity have been examined over the past two decades and in none of them has any trace of abiotically produced organic compounds been found."[8]

Philip H. Abelson of the Carnegie Institution's Geophysical Laboratory declared:

> If the methane-ammonia hypothesis were correct, there should be geochemical evidence supporting it. What is the evidence for a primitive methane-ammonia atmosphere on earth? The answer is that there is no evidence for it, but much against it.[9]

(5) Lightning, which is considerably stronger than Miller's electric sparks, often destroys what it contacts. Lightning, of course, occurs all the time today, and produces nothing resembling elements of life—except in reruns of *Frankenstein* movies.

(6) Miller and his co-author L. E. Orgel wrote that, because organic compounds are unstable, such could probably not have originated life unless the oceans were quite cool—"A temperature of 0°C would have helped greatly," they said.[10] But how could the young Earth, which was supposedly graduating from a hot molten state, have been that cold?

(7) Miller performed his experiment under precise conditions, using a flask as a trap. But ancient oceans had no controlled environment or flasks. Evolution denies that life had a designer. But even if scientists *did* manufacture life in a lab, wouldn't that only show it was an intelligent process?

(8) Perhaps most importantly, the amino acids Miller produced, as we shall see, are a trivial fraction of what life requires.

Amino acids to proteins

Bodies are made up of cells. Cells consist mostly of proteins. And proteins are chains of amino acids. So before we got a cell, Miller's amino acids would first have to form proteins.

A protein must be magnified a million times to become visible to the eye.[11] Human cells use about 200,000 different proteins. Most proteins have hundreds of amino acids.

Every amino acid has at least one activating enzyme. Protein formation requires that amino acids be activated by their enzymes and collected by a substance called *transfer-RNA*. Did Miller's lightning, striking the primordial soup, also produce enzymes and transfer-RNA?

Another difficulty for the Oparin-Miller model: when an amino acid joins a chain, a water molecule is released. Therefore, if water is present, a water molecule will normally be given back, and the chemical process reversed.[12] How, then, could proteins have formed in ancient oceans? As A. E. Wilder-Smith noted:

> The ocean is thus practically the last place on this or any other planet where the proteins of life could be formed spontaneously from amino acids.[13]

Aware of the problem, evolutionists have invented different scenarios, such as that the amino acids washed up on a volcano's rim where heat evaporated the water. But what does that do to the Miller-Orgel thesis of a cool environment? Such fanciful hypotheses have not enjoyed wide acceptance.[14]

Chances are

Poker's rarest hand is a straight flush (five numerically consecutive cards, all of the same suit). Play the game tonight, you're not apt to get one. But what if you played for months on end? Sooner or later, you might land a straight flush.

Or suppose I had tried to bat against Nolan Ryan when the ace pitcher was at his peak. Not much chance for a guy like me to get a hit off Ryan. But what if I could have swung away hundreds or even thousands of times? Eventually, if only by sheer luck, I'd have gotten good wood on one of his pitches, and sent the ball to the outfield for a hit.

Evolution uses this type of argument, asserting that, given enough time, the right amino acids could have assembled together by chance and formed proteins.

Is this true? If you had a deck of cards numbered one to ten, and shuffled them, the odds that they would come out in precise order—one through ten—are only one in 3,628,800. Now, suppose you had a deck (or any other system) consisting of 100 components. The chances of appearing in precise order would be one in 10^{158}, or a one followed by 158 zeroes. Again, anything with odds greater than 1 in 10^{50} is considered effectively impossible.

Our analogy is imperfect, since the body uses only twenty different amino acids; but the average protein chain consists of well over a hundred of these, which must be in exact order, so we start to sense the problem's magnitude. For more precise analysis, I quote Nobel Prize winner Francis Crick, who, along with James Watson, determined DNA's molecular structure:

> If a particular amino acid sequence was selected by chance, how rare an event would this be? . . . Suppose the chain is about two hundred amino acids long; this is, if anything, rather less than the average length of proteins of all types. Since we have just twenty possibilities at each place, the number of possibilities is twenty multiplied by itself some two hundred times. This is conveniently written 20^{200} and is approximately equal to 10^{260}, that is, a one followed by 260 zeros. This number is quite beyond

> our everyday comprehension. . . . The great majority of sequences can never have been synthesized at all, at any time.[15]

Hemoglobin, the blood's oxygen-carrying protein, has two chains, alpha and beta, containing a total of 287 amino acids. Some individuals inherit a gene which causes the amino acid valine to substitute for the amino acid glutamic acid, at position 6 of the beta chain. The result is sickle cell anemia—even though the other 286 amino acids are assembled perfectly. This shows how flawless a protein's structure must be.

Sir Fred Hoyle, the eminent British astronomer, stated:

> At all events, anyone with even a nodding acquaintance with the Rubik cube will concede the near-impossibility of a solution being obtained by a blind person moving the cube faces at random. Now imagine 10^{50} blind persons each with a scrambled Rubik cube, and try to conceive of the chance of them all *simultaneously* arriving at the solved form. You then have the chance of arriving by random shuffling of just one of the many biopolymers [proteins, etc.] on which life depends. The notion that not only the biopolymers but the operating programme of a living cell could be arrived at by chance in a primordial organic soup here on the Earth is evidently nonsense of a high order.[16]

Sir Bernard Lovell, the astronomer who built the world's first completely steerable radio telescope, said:

> The operation of pure chance would mean that within the half billion to a billion year period the organic molecules in the primeval seas might have to undergo 10^{130} trial assemblies in order to hit upon the correct sequence. The possibility of such a *chance* occurrence leading to the formation of one of the smallest protein molecules is unimaginably small. Within the boundary conditions of time and space we are considering it effectively zero.[17]

Zoologist Harold Coffin summed the findings of James Coppedge:

> Coppedge calculates that the odds are 10^{161} to one that not one usable protein would result from chance even if all the atoms on the earth's surface, including water, air, and the crust of the earth were made into conveniently available amino acids and 4 to 5 billion years were involved.[18]

In short, chance could not produce even one protein. And a "simple," single-celled bacterium contains *thousands* of different proteins. What, then, are the odds of assembling *all* the proteins necessary for life? You probably don't want to know. Coffin noted:

> Morowitz has determined the probability for the origin of the organic precursors for the smallest likely living entity by random processes. He

> based his calculations on reaction probabilities, a somewhat different and more accurate approach than most other such computations. The chances for producing the necessary molecules, amino acids, proteins, et cetera, for a cell one tenth the size of the smallest known to man (*Mycoplasm hominis H. 39*) is less than one in $10^{340,000,000}$ or 10 with 340 million zeros after it.[19]

Much of the misconception about the probabilities for life traces back to "Darwin's bulldog," Thomas Huxley. In making his case for chance origins, Huxley reportedly said that six monkeys, poking randomly at typewriters for millions of years, could write all the books in the British Museum. We have all heard variations on this theme, such as typing monkeys recreating the works of Shakespeare.

However, anyone who believes these projections hasn't figured the math. What are the odds of a monkey typing one predetermined nine-letter word, such as "evolution"? Since the alphabet has 26 letters, one must multiply 26 by itself eight times. We find the monkey would need, on average, more than five trillion attempts just to write "evolution" once correctly. Typing ten letters per minute, this would take over a million years. If one word is that hard to get, one begins to fathom the difficulties of randomly producing a paragraph, *Hamlet*—or the components of life.

Creation scientist Duane Gish put the monkey matter in perspective:

> A monkey typing 100 letters every second for five billion years would not have the remotest chance of typing a particular sentence of 100 letters even once without spelling errors.
>
> In fact, if one billion (10^9) planets the size of the earth were covered eyeball-to-eyeball and elbow-to-elbow with monkeys, and each monkey was seated at a typewriter (requiring about 10 square feet for each monkey, of the approximately 10^{16} square feet available on each of the 10^9 planets), and each monkey typed a string of 100 letters every second for five billion years (about 10^{17} seconds) the chances are overwhelming that not one of these monkeys would have typed the sentence correctly! Only 10^{41} tries could be made by all these monkeys in that five billion years. . . . There would not be the slightest chance that a single one of the 10^{24} monkeys (a trillion trillion monkeys) would have typed a preselected sentence of 100 letters (such as "The subject of this *Impact* article is the naturalistic design of life on the earth under assumed primordial conditions") without a spelling error, even once. . . .
>
> Considering an enzyme, then, of 100 amino acids, there would be no possibility whatever that a single molecule could ever have arisen by pure chance on the earth in five billion years.[20]

Some may object that evolution is not just a chance process, that natural selection improves the odds. But natural selection operates only in living things; here we are discussing inanimate chemicals that evolutionists themselves say *preceded* the start of life. No one wins this race for survival—all the contestants are dead at the starting line.

Even if luck did put amino acids and proteins together, would that create a living organism? Chemist John Keosian wrote:

> Let it be assumed that the prebiotic waters contained all of the material that "chemical evolution" is supposed to have brought about. Then how, and in what form, could life have arisen from such a scattered melange? The question must be answered, if there is an answer, to give meaning and direction to the pursuit of chemical evolution, otherwise that pursuit will continue to be an endless series of laboratory experiments unrelated to the central problem. There has been a good deal of uncritical acceptance of experiments, results, and conclusions which we are all too ready to acknowledge because they support preconceived convictions. . . .[21]

We can drop sugar, flour, baking powder, and an egg on the floor—but they won't turn into a cake by themselves. We have to mix and bake them according to a recipe. Throwing steel, rubber, glass and plastic together doesn't make a car. It takes skillful engineering. How much more, then, would intelligence be needed to design life? If brilliant scientists have failed to create it, how could blind stupid chance?

Hard cell

Amino acids . . . proteins . . . the next step would be cells. Green and Goldberger, in their book *Molecular Insights into the Living Process*, noted that "the macromolecule-to-cell transition is a jump of fantastic dimensions, which lies beyond the range of testable hypothesis."[22] A typical cell contains ten million million atoms. If you built a scale model of one using tennis ball-sized "atoms," at one atom per minute, it would take 50 million years.[23] Carl Sagan noted:

> The information content of a simple cell has been established as around 10^{12} bits, comparable to about a hundred million pages of the Encyclopaedia Britannica.[24]

Even Richard Dawkins, one of evolution's most outspoken defenders, acknowledges that a cell nucleus "contains a digitally coded database larger, in information content, than all 30 volumes of the *Encyclopaedia Britannica* put together. And this figure is for *each* cell, not all the cells of a body put together."[25]

Molecular biologist Michael Denton takes his readers on a remarkable tour of a cell's interior:

> To grasp the reality of life as it has been revealed by molecular biology, we must magnify a cell a thousand million times until it is twenty kilometres in diameter and resembles a giant airship large enough to cover a great city like London or New York. What we would then see would be an object of unparalleled complexity and adaptive design. On the surface of the cell we would see millions of openings, like the port holes of a vast space ship, opening and closing to allow a continual stream of materials to flow in and out. If we were to enter one of these openings we would find ourselves in a world of supreme technology and bewildering complexity. We would see endless highly organized corridors and conduits branching in every direction away from the perimeter of the cell, some leading to the central memory bank in the nucleus and others to assembly plants and processing units. The nucleus itself would be a vast spherical chamber more than a kilometre in diameter, resembling a geodesic dome inside of which we would see, all neatly stacked together in ordered arrays, the miles of coiled chains of the DNA molecules. A huge range of products and raw materials would shuttle along all the manifold conduits in a highly ordered fashion to and from all the various assembly plants in the outer regions of the cell. . . .
>
> We would see around us, in every direction we looked, all sorts of robot-like machines. We would notice that the simplest of the functional components of the cell, the protein molecules, were astonishingly, complex pieces of molecular machinery, each one consisting of about three thousand atoms arranged in highly organized 3-D spatial conformation. We would wonder even more as we watched the strangely purposeful activities of these weird molecular machines, particularly when we realized that, despite all our accumulated knowledge of physics and chemistry, the task of designing one such molecular machine—that is one single functional protein molecule—would be completely beyond our capacity at present and will probably not be achieved until at least the beginning of the next century. . . .
>
> We would see that nearly every feature of our own advanced machines had its analogue in the cell: artificial languages and their decoding systems, memory banks for information storage and retrieval, elegant control systems regulating the automated assembly of parts and components, error fail-safe and proof-reading devices utilized for quality control, assembly processes involving the principle of prefabrication and modular construction. In fact, so deep would be the feeling of *deja-vu*, so persuasive the analogy, that much of the terminology we would use to describe this fascinating molecular reality would be borrowed from the world of late twentieth-century technology.[26]

So much for a cell being a "simple little lump of albuminous combination of carbon," as Haeckel put it. The cell has machines more intricate than any made by man. When in the history of science did a machine arise spontaneously?

Furthermore, cells are, as Behe would say, "irreducibly complex." Each cellular structure functions only in coordination with others. Explaining how such things arose independently poses insurmountable difficulties for evolution. For example, cell membranes depend on proteins, yet proteins cannot function without a cell membrane. In *Scientific American*, John Horgan described another closed circle:

> DNA cannot do its work, including forming more DNA, without the help of catalytic proteins, or enzymes. In short, protein cannot form without DNA, but neither can DNA form without proteins.[27]

But wait! DNA? There's another hurdle for Darwinism—*the genetic code*. How do arm cells know they are arm cells, not nose cells? How do the cells of a fetal giraffe know to grow up as a giraffe, not a zebra? They require instructions.

Genes carry hereditary information. *DNA* is the substance of which genes are made. The *genetic code* is a real code, found in DNA molecules, that tells the cells what to do. We may look on a cell as a little factory, proteins as machines in the factory, and the genetic code as instructions for assembly of the machines.[28]

The code is an actual language that scientists have deciphered. It consists of substances called nucleotides, of which there are four. They are distinguished by their bases: (A) adenine, (T) thymine, (G) guanine, and (C) cytosine. These function just like an alphabet. A gene consists of about 1,000 nucleotides, which normally appear in triplets, such as AGC or ATG. Most triplets specify amino acids but some indicate "stop," just as a telegram will say "stop" to end a sentence. All of the world's organisms—animals, plants, and even viruses—depend on this code for existence.

So cells, like computers, are *programmed* for their functions. The genetic code is far more complex than the codes Microsoft engineers used to design Windows 95. Anyone believe Windows 95 could turn up without intelligence? How, then, the genetic code? Of course, evolutionists have their stock answer: time and chance. In other words, if you had a deck of cards, each bearing a letter, and kept shuffling them, eventually you would get a meaningful sentence like "I LOVE YOU."

But A. E. Wilder-Smith pointed out the fallacy of such reasoning.[29] Suppose you handed your "I LOVE YOU" sentence to a man who spoke only German? To have meaning, a message must be received by someone who understands it.

Cells *translate* the genetic code's instructions. Saying that chance produced the code is insufficient. We must also believe chance created the cellular translation devices. How could a translation device, formed by chance, interpret information also formed by chance? How could a haphazard machine translate something with no meaning into meaning? Caryl P. Haskins's comments in *American Scientist* were understated:

> Did the code and the means of translating it appear simultaneously in evolution? It seems almost incredible that any such coincidence could have occurred, given the extraordinary complexities of both sides and the requirement that they be coordinated accurately for survival.[30]

Which came first? Not likely the genetic code, if there was nothing to translate it. That would be like books existing before there were people to read them. But why would translation devices evolve first, if there was no genetic code to read? This is yet another irreducible complexity. Evolutionists Robert Augros and George Stanciu wrote in *The New Biology* (1987):

> What cause is responsible for the origin of the genetic code and directs it to produce animal and plant species? It cannot be matter because of itself matter has no inclination to these forms, any more than it has to the form Poseidon or to the form of a microchip or any other artifact. There must be a cause apart from matter that is able to shape and direct matter. Is there anything in our experience like this? Yes, there is: our own minds. The statue's form originates in the mind of the artist, who then subsequently shapes matter, in the appropriate way. . . . For the same reasons there must be a mind that directs and shapes matter in organic forms. . . . This artist is God and nature is God's handiwork.[31]

Matter is not intrinsically informative. To say that a human cell was built by its chemicals is like saying a book was written by its paper and ink, or that a typewriter was constructed by the iron in its frame. Functional design requires more than matter—it takes intelligence.

The U.S. government has built a number of radio telescopes at great expense. Evolutionists hoped these might detect radio signals from intelligent beings elsewhere in the universe. After all, if chance creates life, more life should be out there. So far, these telescopes have received no signals from other civilizations. But if they did, how would we know the messages weren't produced by chance? Why, the signals would have a pattern—a *code* revealing intellect. Why, then, is a bacterium's *genetic* code, with enough information to fill encyclopedias, disregarded as happenstance? Somethin' wrong with this here picture.

Another trivial question: When and how did cells acquire the ability to reproduce themselves? Chalk up another one for "chance."

Amino acids. Proteins. Cells. The genetic code. Translation devices. Reproductive ability. A lot for luck to accomplish. If you placed bets with an evolutionist on coin tosses, and you flipped thirty straight heads, he'd accuse you of having a two-headed coin. Sixty straight times, he might come after you with tooth and claw. Yet the same evolutionist, without batting an eyelash, will believe that infinitely less probable events took place—to paraphrase Michael Pitman, that the cosmic casino kept turning up straight flushes.

One of my favorite shows back in the seventies was *The Rockford Files*, starring James Garner. One episode was called "The Great Blue Lake Land and Development Company." Rockford battled a crooked real estate firm that was selling parcels of desert land as prime "lakefront property"—assuring customers that, in the future, a lake would be added and the land developed. "Take a look at that!" the salesman would tell his sucker, pointing at barren, wind-swept sand. "There's the shopping center! And, oh! Look over there! There's the country club where you'll be enjoying barbecues every Saturday!"

Evolution's sales pitch is a bit like that. Professor Probability takes us to a plot of empty land with a pond on it, and offers it to us for $40 million.

"$40 million?!" we ask. "But there's nothing here!"

"There will be," winks the professor. "Why, in time, the chemicals in that pond will turn into bacteria, the bacteria into fish, the fish into men, and the men will build a whole civilization. Why, this land will become a city, teeming with hotels and amusement parks. I tell you, my friend, at $40 million, this pond is a bargain!"

Did we really go from "atom to Adam"? Can unthinking molecules turn themselves into thinking human beings? No one has answered that question more picturesquely than Sir Fred Hoyle. *Nature* quoted the astronomer as saying that the probability of higher life forms emerging by chance are comparable to the odds that "a tornado sweeping through a junk-yard might assemble a Boeing 747 from the materials therein."[32]

What, then, is the alternative? I again quote DNA co-discoverer Francis Crick:

> An honest man, armed with all the knowledge available to us now, could only state that in some sense, the origin of life appears at the moment to be almost a miracle, so many are the conditions which would have had to have been satisfied to get it going.[33]

Some, including Hoyle and Crick themselves, have proposed that life was perhaps deposited on Earth from some extraterrestrial source. But that still wouldn't explain how life got started elsewhere! Which leaves us with

God alone. I know—that ires the Darwinist. But as British philosopher G. K. Chesterton said, "It is absurd for the evolutionist to complain that it is unthinkable for an admittedly unthinkable God to make everything out of nothing, and then pretend that it is *more* thinkable that nothing should turn itself into everything."[34]

Like evolution's other branches, theories of origins require as much faith as religion. Anthropologist Loren Eiseley observed: "After having chided the theologian for his reliance on myth and miracle, science found itself in the unenviable position of having to create a mythology of its own: namely, the assumption that what, after long effort, could not be proved to take place today, had, in truth, taken place in the primeval past."[35]

Ernst Chain, the Nobel Prize-winning biochemist, summed it succinctly:

> I have said for years that speculations about the origin of life lead to no useful purpose as even the simplest living system is far too complex to be understood in terms of the extremely primitive chemistry scientists have used in their attempts to explain the unexplainable that happened billions of years ago. God cannot be explained away by such naive thoughts.[36]

PLATE 13. Thomas Huxley, "Darwin's Bulldog"

PLATE 14. Stanley Miller with his apparatus

PLATE 15. To make chance origin of life possible, Soviet biochemist Alexander Oparin proposed that the early Earth had an oxygen-free atmosphere.

Plate 16. Astronomer Fred Hoyle

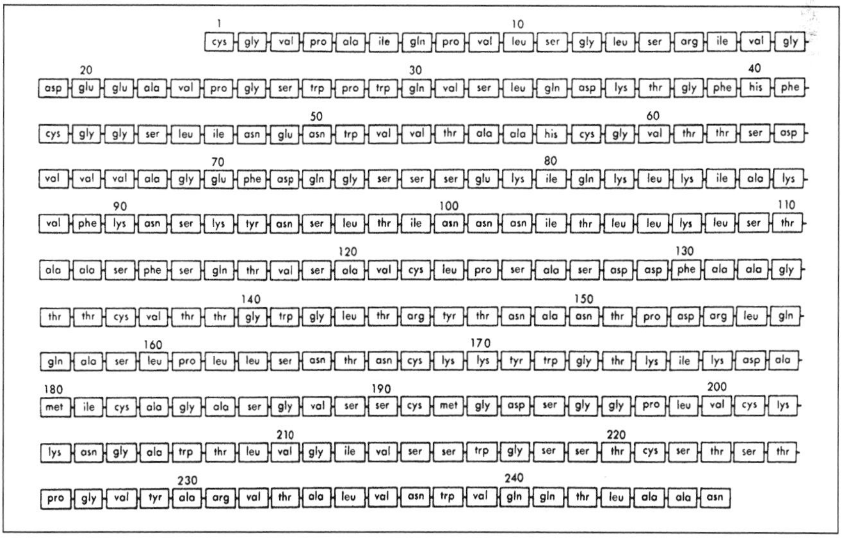

1 10
cys gly val pro ala ile gln pro val leu ser gly leu ser arg ile val gly
20 30 40
asp glu glu ala val pro gly ser trp pro trp gln val ser leu gln asp lys thr gly phe his phe
50 60
cys gly gly ser leu ile asn glu asn trp val val thr ala ala his cys gly val thr thr ser asp
70 80
val val val ala gly glu phe asp gln gly ser ser ser glu lys ile gln lys leu lys ile ala lys
90 100 110
val phe lys asn ser lys tyr asn ser leu thr ile asn asn asn ile thr leu leu lys leu ser thr
120 130
ala ala ser phe ser gln thr val ser ala val cys leu pro ser ala ser asp asp phe ala ala gly
140 150
thr thr cys val thr thr gly trp gly leu thr arg tyr thr asn ala asn thr pro asp arg leu gln
160 170
gln ala ser leu pro leu leu ser asn thr asn cys lys lys tyr trp gly thr lys ile lys asp ala
180 190 200
met ile cys ala gly ala ser gly val ser ser cys met gly asp ser gly gly pro leu val cys lys
210 220
lys asn gly ala trp thr leu val gly ile val ser ser trp gly ser ser thr cys ser thr ser thr
230 240
pro gly val tyr ala arg val thr ala leu val asn trp val gln gln thr leu ala ala asn

Plate 17. Amino acid sequence of the protein chymotrypsinogen demonstrates life's complex precision.

B.C.
BY JOHNNY HART
WHAT DO YOU THINK OF THE HYPOTHESIS: 'WE ALL EVOLVED FROM A ONE-CELLED AMOEBA'?
© Creators Syndicate, Inc., 1987
Dist. by L.A. Times Syndicate
4·29
'HASN'T GOT A LEG TO STAND ON.
hart

PLATE 18

CHAPTER 8

An Ape-man for All Seasons

A few years ago, the story goes, a reporter was interviewing the Chicago Cubs' manager on opening day. "Tell me," he asked, "how many of the players in your lineup would you classify as real sluggers?"

"All nine of them," the manager replied without hesitation.

The reporter looked at him in disbelief. "Oh, really?" he said, barely able to conceal a smirk. "Well, well. Nine sluggers in one lineup! I never heard of that before."

"Neither have I," said the Cubs manager. "I thought you said *sluggards*."

The preceding chapter examined the beginning of the evolutionary time scale. Now we look at its end—our supposed "ape-man" ancestors. A number of candidates have been proposed. We are going to inspect the roster and see if they qualify as sluggers or sluggards.

Man sharing his descent with apes was probably Darwin's most controversial claim. People didn't mind so much the idea that animals evolved from other animals—but putting man in the same category offended many. Darwin's supporters therefore considered discovery of a "missing link" crucial to preserving the entire theory. However, as zoologist Austin H. Clark of the Smithsonian Institution long ago stated:

> Man is not an ape, and in spite of the similarity between them there is not the slightest evidence that man is descended from an ape. . . . From time to time various "missing links" supposed to connect man with the apes have been described. . . . [But] in light of all the evidence available at the present time there is no justification in assuming that such a thing as a "missing link" ever existed, or indeed could have ever existed.[1]

In Chapter Two, we noted that paleontology, the study of fossils, has failed to discover transitional forms for animals. *Paleoanthropology* is the study of fossils of hominids (forms of humans) and other evidences of ancient man. Most of us recall school textbooks with drawings of man's progression from ape, each picture showing the creature less hairy, less brutelike, and walking more upright. It was rarely explained that these gradations were the artist's imagination. Jerold Lowenstein and Adrienne Zihlman noted in *New Scientist*:

> Imaginations run riot in conjuring up an image of our most ancient ancestor—the creature that gave rise to both apes and humans. This ancestor is not apparent in ape or human anatomy nor in the fossil record, but is evident only in the unseen world of the genome within the cell.[2]

John Reader, author of *Missing Links*, said in *New Scientist*:

> [E]ver since Darwin's work inspired the notion that fossils linking modern man and extinct ancestor would provide the most convincing proof of human evolution, preconceptions have led evidence by the nose in the study of fossil man.[3]

Lord Zuckerman, the famed anatomist of the University of Birmingham, remarked:

> For example, no scientist could logically dispute the proposition that man, without having been involved in any act of divine creation, evolved from some ape-like creature in a very short space of time—speaking in geological terms—without leaving any fossil traces of the steps of the transformation.
>
> As I have already implied, students of fossil primates have not been distinguished for caution when working within the logical constraints of their subject. The record is so astonishing that it is legitimate to ask whether much science is yet to be found in this field at all. The story of the Piltdown Man hoax gives a pretty good answer.[4]

The jawed fraud

Between 1908 and 1912, Charles Dawson, an amateur fossil hunter, recovered pieces of an old human skull from a gravel pit in Piltdown, England, a few miles from Darwin's home. The find impressed Arthur Smith Woodward of the British Museum. They dug deeper at Piltdown, joined by Pierre Teilhard de Chardin, a controversial Jesuit priest. (A paleontologist and outspoken Darwinist, Teilhard declared that evolution "is a general condition to which all theories, all hypotheses, all systems must bow and which they must satisfy henceforward if they are to be thinkable and true. Evolution is a light illuminating all facts, a curve that all lines must follow.")[5]

Further digging produced a lower jaw apelike in shape, but with teeth too short for an ape. The skull and lower jaw were assigned to the same individual, who was said to be at least 500,000 years old. Dawson and Woodward announced the find in December 1912, and the world was told that Darwin's "missing link" had been found. Though its editorial page revealed misgivings, the *New York Times* ran a story on it headlined "Darwin Theory Proved True."[6]

For the next four decades, Piltdown Man was evolution's greatest showcase, featured in textbooks and encyclopedias. Many papers and doctoral dissertations were written on Piltdown Man, which was designated *Eoanthropus* ("dawn man") *dawsoni* in Dawson's honor. The scientists who had worked on the fossil and verified its authenticity—Woodward, anatomist Arthur Keith, and brain specialist Grafton Elliot Smith—were all knighted. In the meantime, clergymen who had denounced evolution were ridiculed; Piltdown, it was said, had proven them wrong.

The British Museum displayed a plaster cast of Piltdown Man, and hundreds of thousands gazed upon their ape-man "ancestor." The actual fossils were kept locked up. In 1953, however, scientists subjected them to a new method of chemical analysis, which proved the apelike jawbone very recent. Close inspection also revealed file marks on the teeth—they had been whittled down to make them shorter and more human-looking. And the bones had been treated with chemicals to increase their apparent age. The jaw had clearly been planted, and bore no relation to the human skull, which was determined to date to the Middle Ages but not hundreds of thousands of years ago. (The Piltdown site had been used as a mass grave during the great plague of A.D. 1348–49.) In 1982, collagen testing proved conclusively that the jawbone was an orangutan's.

No one knows for sure who perpetrated the fraud. Investigative books have suggested more than a dozen suspects—even Sherlock Holmes creator Arthur Conan Doyle, who lived near the site! A few writers, including Stephen Jay Gould, have made a case for Pierre Teilhard de Chardin. In 1912, some scientists had expressed doubts about the jawbone's relationship to the skull, saying that if a canine tooth were found, it would resolve the matter. Teilhard then "fortuitously" chanced upon a canine tooth at his feet, while sitting on a gravel pile at the Piltdown site. Investigators later determined Teilhard's tooth had been filed down and painted. But did he plant it, or was it put there for him to find?

Darwinists have tried to save face in the Piltdown matter by saying it shows how evolutionary science "corrects itself." But the rectification took some forty years. And regardless of who committed the fraud, Piltdown proved that even a multitude of experts, blinded by preconceptions, could be deceived. As a result, an entire generation was deceived with them.

The flim-flam ham

America was not to be outdone in the adventurous search for a missing link. In the early 1920s, paleontologist Henry Fairfield Osborn, an ardent evolutionist, was battling William Jennings Bryan over the teaching of evolution in public schools. Bryan, of course, went on to aid the prosecution in the famous Scopes trial. In 1922, geologist Harold Cook discovered a single tooth in Nebraska. After examining it, Osborn declared it belonged to an early ape-man, whom he named *Hesperopithecus haroldcookii* in Cook's honor. Popularly, it became known as "Nebraska Man."

Osborn was particularly delighted because Nebraska was Bryan's home state. He gloated that "the Earth spoke to Bryan and spoke from his own State of Nebraska, in the message of a diminutive tooth, the herald of anthropoid apes in America. . . . this little tooth speaks volumes of truth—truth consistent with all we have known before, with all that we have found elsewhere."[7]

William K. Gregory and Milo Hellman, specialists in teeth at the American Museum of Natural History, said after careful study that the tooth was from a species closer to man than ape.[8] Harris Hawthorne Wilder, a zoology professor at Smith College, wrote in his book *The Pedigree of the Human Race*:

> Judging from the tooth alone the animal seems to have been about halfway between *Pithecanthropus* [Java Man, discussed next] and the man of the present day, or perhaps better between *Pithecanthropus* and the man of the Neandertal type. . . .[9]

In England, Grafton Elliot Smith, who had been involved in the Piltdown affair, convinced *The Illustrated London News* to publish an artist's rendering of Nebraska Man. The picture, which appeared in a two-page spread and received wide distribution, showed two brutish, naked ape-persons, the male with a club, the female gathering roots. All this from one tooth.

However, further excavations at Cook's site revealed that the tooth belonged neither to ape nor man, but to a peccary, a close relative of the pig. "Nebraska Man" was unceremoniously withdrawn as a "missing link."

Dubious Dubois

When I studied biology in high school, our textbook featured a beetle-browed ape-man called "Java Man." We were assured that this moron was our ancestor. G. K. Chesterton aptly commented on Java man's depiction in his time:

> A detailed drawing was reproduced carefully shaded to show the very hairs of his head were all numbered. No uninformed person, looking at its carefully lined face, would imagine for a moment that this was the portrait of a thigh bone, of a few teeth and a fragment of a cranium.[10]

The Java Man story actually begins with Ernst Haeckel, the German zoologist who made those imaginative drawings of "spontaneously generated" bacteria. Haeckel was convinced that ape-men must have existed, and—again without evidence—he commissioned a painting of such, whom he named *Pithecanthropus alalus*: ape-man without speech.

One of Haeckel's students, Eugene Dubois, became determined to find *Pithecanthropus*. This was before Piltdown. Glory and honor were sure to cover the first man who could validate Darwinism by finding the missing link. Haeckel believed men had separated from apes somewhere in Africa or Southern Asia.

In 1887, Dubois signed up as a doctor for eight years with the Dutch medical corps in the Dutch East Indies (now Indonesia), intending to hunt for fossils during all his spare time. The authorities there permitted him to search; he was given fifty forced laborers and two army corporals as supervisors.

Years of excavation produced little of significance. Then, in 1891, along Java's Solo River, the laborers dug up a tooth and a skullcap. The latter was apelike, having a low forehead and large eyebrow ridges. The following year, about forty feet away, the diggers unearthed a thigh bone that was clearly human. As naturalist Richard Carrington related:

> Dubois was at first inclined to regard his skull cap and teeth as belonging to a chimpanzee, in spite of the fact that there is no known evidence that this ape or any of its ancestors ever lived in Asia. But on reflection, and after corresponding with the great Ernst Haeckel, Professor of Zoology at the University of Jena, he declared them to belong to a creature which seemed admirably suited to the role of the "missing link."[11]

Dubois, like Piltdown's discoverers, presumed that an apelike bone somewhere near a human bone meant the two belonged to the same creature, constituting Darwin's link. Following Haeckel's lead, Dubois named his find *Pithecanthropus erectus*, "erect ape-man." Haeckel, who had not even seen the bones, telegraphed Dubois: "From the inventor of Pithecanthropus to his happy discoverer!"[12]

In 1895, Dubois returned to Europe to make his case. He went on a lecture circuit and displayed his fossils at the International Congress of Zoology in the Netherlands. The response from experts was mixed, however. Rudolph Virchow, who had once been Haeckel's professor and is regarded as the father of modern pathology, said: "In my opinion this creature was an animal, a giant gibbon, in fact. The thigh bone has not the slightest connection with the skull."[13]

The circumstances of Dubois' find were unorthodox. He had apparently been absent when the laborers dug up his fossils.[14] Maps and diagrams of the site were not made until after the excavation. Under such conditions, a modern dig would be disregarded.

Although he had studied anatomy and medicine, Dubois had no formal training, at the time, in geology or paleontology. Little was known about Java's terrain, and he was unqualified to date the layers containing the fossils. In fact, after making the discovery, he relabeled them—making his ape-man older. G. H. R. von Koenigswald, a paleoanthropologist in Java for many years, remarked:

> When Dubois issued his first description of the fossil Javanese fauna he designated it Pleistocene. But no sooner had he discovered his *Pithecanthropus* than the fauna had suddenly to become Tertiary. He did everything in his power to diminish the Pleistocene character of the fauna. . . .[15]

Angered by the lukewarm reception from scientists, Dubois became secretive and would not let anyone else inspect the bones. While vigorously defending his find as uniquely constituting the missing link, he sharply criticized fossils found by others. His egotism even rankled other evolutionists. As von Koenigswald related:

> *Pithecanthropus* became Dubois' destiny. It was his discovery, his creation, his exclusive possession; on this point he was as unaccountable as a jealous lover. Anyone who disagreed with his interpretation of *Pithecanthropus* was his personal enemy. When his ideas failed to win general acceptance he sullenly withdrew, growing mistrustful, unsociable and eccentric. . . . at night he used to hear burglars prowling round the house, bent on stealing his *Pithecanthropus.*[16]

In 1907, an expedition of German scientists from various disciplines, led by Professor M. Lenore Selenka, traveled to Java seeking more clues to man's ancestry in the region of Dubois' discovery. Typically, Dubois did not cooperate, refusing to let them see his precious bones. (Secrecy—the hallmark of true science.)

The expedition established a barracks, hired 75 workers, and methodically removed 10,000 cubic meters of material. It sent 43 crates of fossil material back to Germany. However, no evidence for *Pithecanthropus* was found. In the stratum of Dubois' find, the scientists found hearths, and flora and fauna that looked rather modern. The expedition's report also noted a nearby volcano that caused periodic flooding in the area. Java Man had been found in volcanic sediments. The report observed that the chemical nature of those sediments, not ancient age, probably caused the fossilization of *Pithecanthropus*. And native tradition indicated the Solo River had changed course in the thirteenth or fourteenth century; if true, Dubois' fossils might not be much over 500 years old. Dr. E. Carthaus, a geologist on the expedition, concluded that *Pithecanthropus* was contemporary with modern humans.[17]

Raising further questions: Dubois' own excavations had also recovered two skulls, clearly human, from another site called Wadjak. However, he declined to display them when trumpeting Java Man. In fact, he kept the Wadjak skulls hidden under the floorboards of his house for thirty years.[18] The reason is speculative, but perhaps Dubois believed that if anyone suspected Java Man had coexisted with modern humans, then it could not be the missing link.

Why did Dubois finally reveal the skulls in 1920? Sir Arthur Keith wrote: "The event which brought them from Dr. Dubois' fossil cupboard was the publication of Dr. Stewart A. Smith's monograph on the Talgai man—an inhabitant of Australia in Pleistocene times."[19] With his Java Man fame fading, Dubois apparently could not resist disclosing the Wadjak skulls, insisting they proved that he, not Smith, had found the first "proto-Australian."

Over the years, fossil fragments similar to Dubois' skullcap were recovered from Java. Since 1950, paleoanthropologists have been calling Java Man *Homo erectus*, an early form of man. But late in life, Dubois, still trying to distance his find from other discoveries, insisted that "*Pithecanthropus* was not a man, but a gigantic genus allied to the gibbons"[20]—a bit ironic, for, although he still called it "the real 'missing link,'" a giant gibbon was just what Virchow had originally dismissed Java Man as.

Today, the Selenka findings and various deficiencies of Dubois' work are largely ignored, and Java Man remains in textbooks as one of evolution's undisputed "facts."

Bull in the China shop?

Another imbecile ancestor in my high school biology textbook was "Peking Man." In 1919, Dr. Davidson Black, who had worked with Grafton Elliot Smith on the Piltdown finds, traveled to China hoping to discover a missing link of his own. This went unfulfilled until 1927, when he was shown a single tooth. Black announced that it came from an ape-man, whom he called *Sinanthropus pekinensis*—a speculation that rivaled Nebraska Man's for boldness.

The Rockefeller Foundation provided a large grant to establish a laboratory in China to continue the search. Black and his team continued digging at the same site—a limestone hill, containing a collapsed cave, called Chou Kou Tien ("Dragon Bone Hill"). The excavation lasted until World War II began, and turned up fragments of 14 skulls, 11 jawbones, and 147 teeth. Numerous animal fossils were also recovered.

Davidson Black died in 1934, but not before he received many honors for discovering *Sinanthropus pekinensis,* including membership in Britain's Royal Society. He was replaced by Franz Weidenreich, who, like Osborn and Black, was easily smitten by the tooth fairy. Based upon some large molar teeth bought in Hong Kong apothecary shops, Weidenreich described yet another missing link: *Gigantopithecus*—a creature who, he claimed, was twice the size of a male gorilla.[21] *Gigantopithecus* failed to catch fire with evolutionists, however—as William Fix notes, "This ancestral model had a shorter run than the Edsel."[22]

Another major figure at the Chou Kou Tien excavations was Pierre Teilhard de Chardin, one of the prime Piltdown suspects. The Vatican had banished Teilhard to China in 1926 for his unorthodox religious views and advocacy of evolution. He had a remarkable proclivity for being around "missing links" at the time of their discovery. Teilhard invited his old professor, Marcellin Boule, to evaluate the Peking Man fossils. Boule headed the Institute of Human Paleontology at the Museum of Natural History in Paris. Boule and his co-author H. V. Vallois later noted:

> Black, who had felt justified in forging the term *Sinanthropus* to designate *one* tooth, was naturally concerned to legitimize this creation when he had to describe a skull-cap.[23]

Oddly, although skulls, jawbones and teeth were found at the site, there were hardly any fragments of other body parts. As Dunbar and Waage observed in *Historical Geology*: "The absence or rarity of other skeletal parts, as well as evidences that the base of each skull had its base broken away in a definite manner, suggests strongly that the heads had been severed from the bodies and that the brains had been eaten."[24] Although the classic paintings of Peking Man show him living in the cave, it is apparent that the heads had been cut off somewhere else and brought to the site. In parts of Asia, monkey brains have long been considered a delicacy. Some have asked: Was Peking Man an ape hunted and killed for food? That his broken bones were mixed with those of many other animals adds some credence to this possibility.

Initially, *Sinanthropus* was deemed a very ancient ape-man. In 1931, in the archaeological journal *Antiquity*, Grafton Elliot Smith made glowing comparisons between the Peking fossils, the fraudulent Piltdown Man, and the scant fragments of Java Man. He then stated:

> It is a very significant phenomenon that at Chou Kou Tien, in spite of the most careful search in the caves during the last three years, no trace whatever of implements of any sort has been found. When it is considered how vast a quantity of fossil remains has been found and the scrupulous care which has been exercised in the search, it must be something more than a mere coincidence that no trace of any stone implements has been found. . . . Those who have been searching in vain for evidence of human craftsmanship on this site are being forced to the conclusion that Peking Man was in such an early phase of development as not yet to have begun to shape implements of stone for the ordinary needs of his daily life.[25]

This was sharply contradicted when Stone Age expert Henri Breuil visited Chou Kou Tien. His 1932 report in the French Journal *L'Anthropologie* revealed thousands of tools at the site, made of stone, bone and quartz—simple chisels, anvils, tools for making tools, etc. He also found a black, cindery deposit seven meters thick, indicating fires had been maintained there for long periods. Clearly, the site had been the locale of a type of human industry. Breuil wrote:

> Many distinguished experts, independently of each other, have expressed to me the thought that a creature, so physically distant from Man, even Neanderthal, could not have been capable of the behavior I have described. In this case, the remains of *Sinanthropus* could be considered as simple hunting trophies, attributable, like the traces of fire and industry, to a true man whose remains we have not yet found.[26]

Boule and Vallois expressed a similar doubt:

> We may therefore ask ourselves whether or not it is over-bold to consider *Sinanthropus* the monarch of Chou Kou Tien when he appears in its deposit only in the guise of a mere hunter's prey, on a par with the animals by which he is accompanied.[27]

It is rarely mentioned, but in 1933, *modern* human skeletons were discovered at Chou Kou Tien. Since these were in the "upper cave," they have been dismissed as comparatively recent. However, throughout the lower excavation layers, the bones of animals, as well as those of Peking Man, were uniformly mixed—revealing no evolutionary progression.[28] Was it, perhaps, these humans who used the tools and fire, and hunted and decapitated *Sinanthropus*?

If so, who was *Sinanthropus*? A primitive ape-man? It seems unlikely that such would have lived side-by-side with modern men (making the latter cannibals). Just a hunted ape? Not according to the semi-human reconstructions of Peking man, begun by Black and completed by Weidenreich. These are the basis of the famous Peking Man portraits seen around the world. The most famous model, "Nelly," actually consisted of fragments from several individuals, the jawbone found some 80 feet from the skull. Where Weidenreich had no fragments, he filled in the gaps with plaster and his judgement. The final image of "Nelly" was completed by Lucille Swann, an American sculptress living in Peking at the time.[29]

Can we always trust reconstructions? Earnest Albert Hooton, the eminent Harvard anthropologist, wrote: "You can with equal facility model on a Neanderthaloid skull the features of a chimpanzee or the lineaments of a philosopher. These alleged restorations of ancient types of man have very little if any scientific value, and are likely to mislead the public."[30]

Due to the war, the original Peking Man fossils were ordered shipped for safekeeping. *They all then vanished, except for two teeth.* They are gone to this day, and no one knows what became of them. There were many rumors, all uncorroborated: they were lost when invading Japanese troops ransacked a train they were on; they were secretly shipped to Tokyo; they were on a barge that capsized; they were stolen by Chinese dock workers; they were stolen by U.S. marines. In 1951, Wen-chung Pei (who had overseen the excavations) even charged—probably with some prodding from the new Communist government—that the fossils were being secretly held by the American Museum of Natural History.

Whatever happened, the loss was convenient for Darwinism, as there is now no way to subject them to the kind of checking that doomed Piltdown Man, or to definitely verify the reconstructions' accuracy. Evolutionists object that photographs and diagrams still exist, but that is a far cry from having the originals. If all that remained of Piltdown Man was a few pictures, perhaps we'd still be calling him our ancestor.

To balance the perspective, let me say there is no evidence that Weidenreich faked his reconstructions. In fact, when a small piece of bone was found many years later, it reportedly fit perfectly into one of his casts.[31] However, if scientists at the British Museum were fooled into assembling Piltdown Man from a human skull and orangutan jaw, could not preconceptions have also guided Black and Weidenreich? Without knowing what a

Sinanthropus looked like, how does one place bone fragments in their "normal relation"?

Peking Man remains clouded with doubt.

The moot brute

No ape-men review would be complete without discussing Neanderthals. Parts of the first such skeleton were recovered from a quarry in Germany's Neander valley ("thal" or "tal" means "valley"). The unusual-looking bones were brought to the noted pathologist Rudolph Virchow at the University of Berlin. Virchow said they belonged to a man suffering from arthritis and rickets (a bone disease caused by vitamin D deficiency), as well as blows to the head.

But Thomas Huxley, "Darwin's bulldog," saw them as evidence of a missing link. This view became especially prominent as similar bones were unearthed at other sites. Marcellin Boule originated the classic description of Neanderthals. Working with a fairly complete skeleton found near the village of La Chapelle-aux-Saints in 1908, Boule depicted the Neanderthal as an ape-man, stooping with head thrust forward and knees bent, suggesting an ape-like gait. For more than forty years, Neanderthals were presented this way to the public—apish brutes, naked, hairy, and wielding clubs.

However, in 1955, anatomists William J. Straus of Johns Hopkins University and A. J. E. Cave of St. Bartholomew's Hospital Medical College (London) were attending an anatomy conference in Paris. Looking at Boule's reconstruction of the La Chapelle-aux-Saints skeleton, they immediately spotted significant errors, which they summarized in the *Quarterly Review of Biology*. They noted:

> [T]here is nothing in this total morphological pattern to justify the common assumption that Neanderthal man was other than a fully erect biped when standing and walking. It may be that the arthritic "old man" of La Chapelle-aux-Saints, the postural prototype of Neanderthal man, did actually stand and walk with something of a pathological kyphosis [hunchback]; but, if so, he has his counterparts in modern men similarly afflicted with spinal osteoarthritis. He cannot, in view of his manifest pathology, be used to provide us with a reliable picture of a healthy, normal Neanderthalian. Notwithstanding, if he could be reincarnated and placed in a New York subway—provided that he were bathed, shaved, and dressed in modern clothing—it is doubtful whether he would attract any more attention than some of its other denizens.[32]

This skeleton, like the one Virchow examined, had been arthritic. Boule, led by Darwinian suppositions, mistook a hunchbacked condition for an ape-man in the transition of becoming upright.

Another snag: Neanderthal skulls were larger than those of modern humans. This flew in the face of evolutionary tradition, which said that man evolved progressively from creatures with smaller brains and skulls.

Skeletons of many Neanderthals have now been recovered, and they are no longer depicted as ape-men. They are classed along with modern man as *Homo sapiens*. Far from being brutes, Neanderthals buried their dead with great care. One was found interred with evidence of flower bouquets;[33] another with what appeared to be a primitive flute.[34] There is no evidence that Neanderthals were hairy or walked around naked. In 1908, *Nature* related the discovery of a Neanderthal buried in a suit of armor,[35] although evolutionists dismiss the report because it can no longer be verified.

To be sure, Neanderthal skeletons differ from average modern humans. Like Virchow, some believe they suffered from rickets and other bone diseases. Francis Ivanhoe made this case in a 1970 *Nature* article entitled "Was Virchow Right About Neanderthal?"[36] Rickets is not commonly seen today—in America, the disease was almost wiped out by fortifying milk with vitamin D. But in the past, particularly during the ice age, it may have been common, due to poor diet and less sunlight (which the body utilizes to make vitamin D). Rickets produces symptoms resembling Neanderthal traits, especially if a person is afflicted in childhood. However, some observers note that rickets is not evident in all Neanderthal skeletons, and that many Neanderthals appear too robust for bone disease to explain their differences from modern man. Neanderthals may have been a distinct race that became extinct. But in no event were they "ape-men."

By the 1950s, human paleontology was suffering from a bad case of the blahs. Piltdown Man had been exposed as a fraud; Nebraska Man was a pig's tooth; Java Man's discoverer had called him a genus allied to the gibbons; Peking Man's fossils had flown the coop; and Neanderthals had turned out to be *Homo sapiens*. Darwinism desperately needed a shot in the arm from a new ape-man . . .

PLATE 19. The picture of *Pithecanthropus* commissioned by Ernst Haeckel

PLATE 20. Eugene Dubois, discoverer of Java Man

PLATE 21. Sculpture of Piltdown Man, whose actual jaw was an orangutan's, demonstrates the uncertainty of hominid reconstructions.

PLATE 22. *Hesperopithecus*—Nebraska Man—was depicted as above, based on one tooth.

PLATE 23. Piltdown discussion. Piltdown demonstrated how preconceptions could steer leading scientists into error. Rear, from left: British Museum model maker Frank Barlow; neurologist Grafton Elliot Smith; Charles Dawson, who discovered the original fossils; British Museum geologist Arthur Smith Woodward. Front: Arthur Underwood, anatomist Arthur Keith, zoologist William Pycraft, zoologist Edwin Lankester.

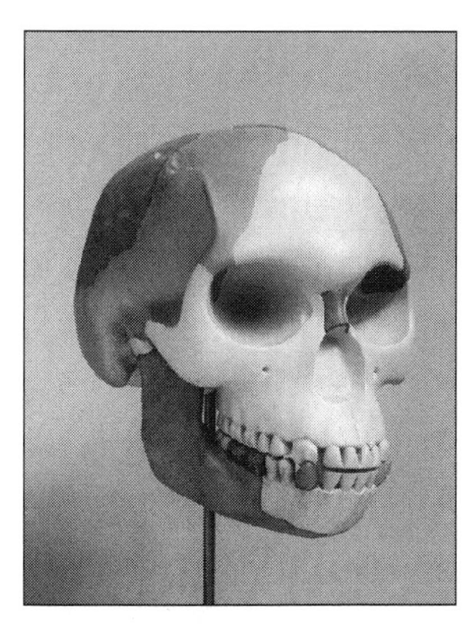

PLATE 24. Reconstruction of the Piltdown skull

Top ten sayings attributed to the Piltdown Man:

(10) "I'm surrounded by idiots."
(9) "Appearances can be deceiving."
(8) "There's a sucker born every minute."
(7) "Nobody makes a monkey out of me!"
(6) "Stupid is as stupid does."
(5) "You ought to have your head examined."
(4) "Never give a sucker an even fake."
(3) "I have a bone to pick with you."
(2) "Let's put our heads together."
(1) "Never look a gift fossil in the mouth."

(Few know it, but prior to his downfall in 1953, Piltdown Man—or "Milty from Pilty" as his fans knew him—was one of Britain's leading stand-up comics, frequently appearing in pubs with anatomist Sir Arthur Keith as his straight man. It was reported that amidst his "Fossil Follies" revue at the London Palladium, Eugene Dubois stormed out during the wisecracking skull's impersonation of Java Man.)

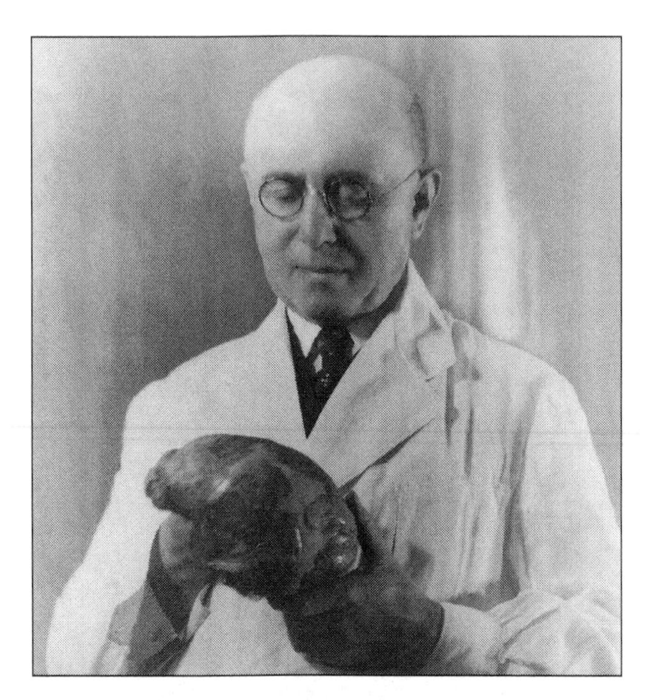

PLATE 25. Franz Weidenreich with *Sinanthropus* skull

PLATE 26. Pierre Teilhard de Chardin with *Sinanthropus* skull

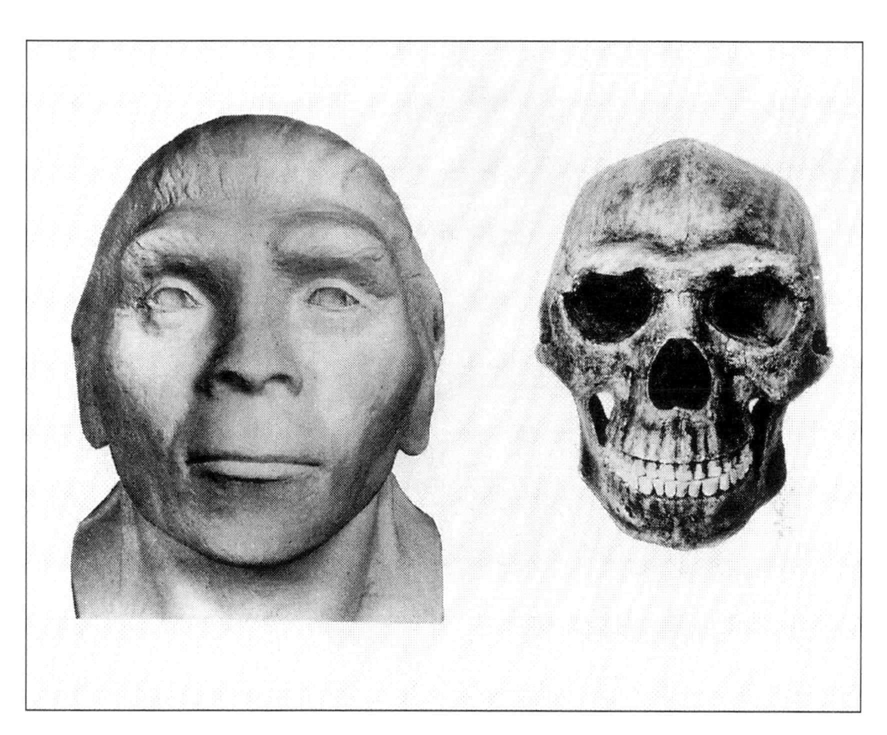

PLATE 27. Reconstructions of head
and skull of "Nelly"

PLATE 28. Early portrayal of Neanderthals

CHAPTER 9

The Reigning World Chimp

Like Davidson Black of "Peking Man" fame, Raymond Dart studied under Grafton Elliot Smith and became a convinced evolutionist. Like Black and Dubois, Dart ventured to a remote place and returned with a "missing link." (If a person was determined to find the missing link in those days, it seems he was likely to find it. As Lowenstein and Zihlman noted in *New Scientist*: "In the course of the past century, the discoverer of every new hominid or hominoid has nominated it as a potential human ancestor.")[1]

Dart went to South Africa, where he was soon led to a small fossilized face and jaw in a cave in the Taung limestone quarry. Convinced the bones had both human and apelike characteristics, Dart announced he had found a missing link, which he called *Australopithecus africanus*, "ape of southern Africa."

Because Dart's discovery coincided with the Scopes trial, his "Taung Child" received favorable press. However, most scientists rejected the find as a young ape[2] (bones of young apes tend to resemble humans more than adults).

But the void created by the Piltdown scandal helped bring the Taung fossil back into vogue, especially as more bones turned up in Africa. Today, australopithecines are widely promoted as the ancestor linking us to apes.

The Taung fossil was initially deemed 2–3 million years old. But in 1973, geologist T. C. Partridge reported in *Nature* that the cave site could not have formed more than 870,000 years ago.[3] This sharply contradicted previous attempts to classify the fossil in man's lineage. However, the Taung Child's difficulties were brushed from the limelight with the discovery of "Lucy" in 1974.

"Lucy" and other australopithecine fossils have been greatly popularized by the National Geographic Society, which initially brought fame and funding to anthropologist Louis Leakey, his wife Mary, and son Richard.

Louis Leakey gained acclaim in 1959, when he found a skull in Tanzania's Olduvai Gorge. Calling it the oldest man ever found, Leakey named it *Zinjanthropus* (East Africa Man), although it has since been reclassified as an australopithecine. Radiometric dating (a subject we will discuss later) gave the fossil an age of 1.75 million years.

Raising questions, however, was a perfectly human skeleton that German anthropologist Hans Reck had long before recovered at a level just above Leakey's find. The Reck fossil has been radiometrically dated at about 17,000 years old.[4] On careful inspection, Reck found no evidence that the skeleton had been intrusively buried—the overlying soil was undisturbed. However, there were protests from the evolutionary community: modern humans weren't supposed to be in old deposits; various explanations were advanced. Reck maintained his position for two decades. Ultimately, he acquiesced and cosigned a letter saying the skeleton was probably an intrusive burial. One wonders, however, to what extent this was compelled by peer pressure in a world of evolutionary preconceptions.

Another discrepancy arose when Mary Leakey, excavating lower than her husband's find, discovered a circular stone structure, clearly the result of purposeful work, resembling shelters still built today by some African tribesmen. How could ape-like australopithecines have produced it?

A further challenge arose in 1978, when footprints were discovered at Laetoli, another Tanzanian site, in a geologic stratum dated 3.5 million years old. Tim White wrote of these in *Science*: "Portions of the trails are eroded but several intact prints are preserved. The uneroded footprints show a total morphological pattern like that seen in modern humans."[5] Russell H. Tuttle, a University of Chicago specialist who studied the footprints extensively, noted that they were "indistinguishable from those of habitually barefoot *Homo sapiens*."[6] How could australopithecines leave modern footprints?

Creation scientist Duane Gish describes with irony *National Geographic*'s depiction of prehistoric Laetoli:

> Footprints of antelopes, pigs, giraffes, elephants, rhinos, hares, ostriches, and other animals were found at Laetoli. In artists' conceptions of the scene, we see pictures of giraffes for the giraffe footprints, elephants for the elephant footprints, ostriches for the ostrich footprints, etc. And—humans for the human footprints? Oh, no! Occupying the human footprints we see a sub-human creature, half-ape and half-man.[7]

National Geographic also brought into prominence the Leakeys' son Richard, with his 1972 discovery of "skull 1470" in Kenya. The fossil looked relatively modern, but Leakey dated it nearly three million years old, and boldly announced:

> Either we toss out this skull, or we toss out our theories of early man. It simply fits no previous models of human beginnings. . . .[It] leaves in ruins the notion that all early fossils can be arranged in an orderly sequence of evolutionary change.[8]

National Geographic "helped" Leakey's claim for the skull's great age by having an artist flesh in the face with an apelike nose, even though nasal bones had not been found (noses, made of cartilage, normally don't fossilize). However, the skull's morphology was too modern for many to accept, and radiometric techniques gave varying dates. Eventually, a consensus was reached for an age under two million years, and Leakey lost his status of finding the world's oldest hominid.

If a missing link could bring its discoverer from obscurity to fame in Piltdown days, it is no less true today. Richard Leakey was upstaged when "Lucy," an australopithecine fossil, was found near Hadar in Ethiopia by Donald Johanson, curator of physical anthropology at the Cleveland Natural History Museum. Initially, Johanson found a knee joint; looking it over, he declared it belonged to a hominid. Based on animal fossils in the area, he considered it three million years old. After recovering other bones, which Johanson also thought were from hominids, his expedition made a press announcement rivaling Leakey's for boldness:

> These specimens clearly exhibit traits which must be considered as indicative of the genus *Homo.* Taken together, they represent the most complete remains of this genus from anywhere in the world at a very ancient time.
>
> All previous theories of the origin of the lineage which leads to modern man must now be totally revised. We must throw out many theories and consider the possibility that man's origins go back to well over four million years.[9]

Johanson's best find was a creature that had been about three and a half feet tall. Sixty percent of the skeleton, including most of the skull, was missing. It was named "Lucy" because the Beatles tune "Lucy in the Sky with Diamonds" was blaring at Johanson's camp at the time. This discovery was also popularized in *National Geographic*, which didn't hesitate to publish a painting of these ancient ancestors, even though the remains had been fragmentary.

By calling his skeleton "Lucy," and the collective fossils from the area "First Family," Johanson helped engineer a popular impression that they were early humans. He began receiving grant money from many sources.

There are difficulties, however, with asserting that Lucy, or any australopithecine, was our ancestor. Australopithecine fossils show that they had long forearms and short hind legs, like today's apes. They also had curved fingers and long curved toes, like those apes use for tree-swinging. Stern and Susman, writing in the *American Journal of Physical Anthropology,* noted:

> There is no evidence that any extant primate has long, curved, heavily muscled hands and feet for any purpose other than to meet the demands of full or part-time arboreal [tree-dwelling] life.[10]

The main substance to the claim that Lucy walked upright was the appearance of the leg bones and hip. Stern and Susman, while agreeing the Hadar finds were bipedal to some degree, found them substantially arboreal:

> In summary, the knee of the small Hadar hominid shares with other australopithecines a marked obliquity of the femoral shaft relative to the bicondylar plane, but in all other respects it falls either outside the range of modern human variation, or barely within it. . . . and since many of these traits may not serve to specify the precise nature of the bipedality that was practiced, we must agree with Tardieu that the overall structure of the knee is compatible with a significant degree of arboreal locomotion.[11]

Britain's Lord Solly Zuckerman, who was raised to peerage for his scientific achievements, was a leading authority on australopithecines, having subjected them to years of biometric testing. Although the following comments were made prior to Lucy's discovery, they are worth noting:

> For my own part, the anatomical basis for the claim that the australopithecines walked and ran upright like man is so much more flimsy than the evidence which points to the conclusion that their gait was some variant of what one sees in subhuman primates, that it remains unacceptable.[12]

Charles Oxnard, former director of graduate studies and professor of anatomy at the University of Southern California Medical School, subjected australopithecine fossils to extensive computer analysis. Stephen Jay Gould called him "our leading expert on the quantitative study of skeletons."[13] Oxnard declared:

> [T]he australopithecines known over the last several decades from Olduvai and Sterkfontein, Kromdraai and Makapansgat, are now irrevocably removed from a place in the evolution of human bipedalism, possibly from a place in a group any closer to humans than to African apes and certainly from any place in the direct human lineage. All of this should

> make us wonder about the usual presentation of human evolution in introductory textbooks, in encyclopaedias and in popular publications. In such volumes not only are australopithecines described as being of known bodily size and shape, but as possessing such abilities as bipedality and tool-using and -making and such developments as the use of fire and specific social structures. Even facial features are happily (and non-scientifically) reconstructed.[14]

He stated in 1987:

> The various australopithecines are, indeed, more different from both African apes and humans in most features than these latter are from each other. Part of the basis of this acceptance has been the fact that even opposing investigators have found these large differences as they too, used techniques and research designs that were less biased by prior notions as to what the fossils might have been. . . . also, most of the new studies have come from laboratories independent of those representing individuals who have found the fossils.[15]

To be sure, many evolutionary anthropologists have called the australopithecines bipedal ancestors of man. But contrary to the general public impression, this view is far from uncontested.

The assumption that two-footed mobility establishes human kinship is groundless. Gorillas occasionally walk bipedally;[16] Tanzanian chimpanzees are seen standing on two legs when gathering fruit from small trees;[17] Zaire's pygmy chimpanzee walks upright so often that it has been dubbed "a living link."[18] *Science News* reports of the latter: "Like modern gorillas they tend to be knuckle-walkers on the ground, yet they seem to be natural bipeds, too, frequently walking upright both on the ground and in the trees."[19] So even if australopithecines did have some limited ability to go on two feet, it doesn't make them man's ancestor any more than these modern apes. For that matter, birds are bipedal—therefore human?

Further complicating the case for australopithecines are out-of-place fossils. An elbow bone was discovered at Kanapoi in Kenya, entombed in a geologic stratum lower than where australopithecines are found. The fossil has been dated 4.5 million years old. The problem? It appears perfectly modern. Based on computer analysis, Henry M. McHenry of the University of California, Davis, stated: "The results show that the Kanapoi specimen, which is 4 to 4.5 million years old, is indistinguishable from modern Homo Sapiens. . . ."[20] So, did apes evolve into man, then into apes again, then back into man?

Other fossils contradict evolutionary theory. In 1866, a modern skull was found in a California gold mine in Pliocene deposits, making it over "two million years old." Sir Arthur Keith said of the discovery:

> The story of the Calaveras skull, although grown stale from frequent repetition, cannot be passed over. It is the "bogey" which haunts the student of early man—repelling some, fascinating others, and taxing the powers of belief of every expert almost to the breaking point. . . . Indeed were such discoveries in accordance with our expectations, if they were in harmony with the theories we have found regarding the date of man's evolution, no one would ever dream of doubting them, much less of rejecting them.[21]

But since the Calaveras skull does not concur with their assumptions, evolutionists reject it as a joke planted by miners.

In 1860, the Italian geologist Raggozoni dug a modern human skeleton out of Pliocene strata, and examination of the overlying rocks showed they were undisturbed (i.e., it was not an "intrusive burial"). Keith wrote: "As the student of prehistoric man reads and studies the records of the 'Castenedolo' finds, a feeling of incredulity rises within him. He cannot reject this discovery as false without doing injury to his sense of truth, and he cannot accept it as fact without shattering his accepted beliefs."[22]

Many other examples can be cited (see, for example, Malcolm Bowden's book, *Ape-men: Fact or Fallacy?* pp. 64–77). Some finds are more controversial than others, but always disbelieved by Darwinists. Harvard anthropologist Earnest Hooton noted: "Heretical and non-conforming fossil men were banished to the limbo of dark museum cupboards, forgotten or even destroyed."[23] Anthropologist G. W. H. Schepers acknowledged:

> And when someone produces relics of *Homo sapiens* in geological deposits more ancient than the Mid Pleistocene, we seek all manner of unlikely explanations for such an "impossibility," even going so far as to discredit usually reliable witnesses. Such finds ultimately become veritable skeletons in the closet to anthropologists, who, in their subconscious endeavor to support dogma even fail to describe such finds fully enough to allow fools to enter where angels fear to tread![24]

Ironically, evolutionists, who once envisioned a whole race of ape-men based on a single tooth—which turned out to be a pig's—will dismiss an entire skeleton when it clashes with Darwinian preconceptions. The truth is, there are limits to how much one can deduce from a bone. Molecular biologist Michael Denton notes that "ninety-nine percent of the biology of any organism resides in its soft anatomy, which is inaccessible in a fossil."[25]

Paleoanthropologists use cranial capacity (skull size) to judge the evolutionary status of our supposed ancestors, but even in modern humans, adult cranial capacity ranges from 700 to 2200 cubic centimeters,[26] and has no bearing on intelligence. People's bone structure greatly varies, based on heredity, age, sex, health, and climate. Some are big-boned, some small-boned. There are Sumo wrestlers and pygmies. Doubtless, our ancient forebears were also diverse in their looks. How, then, can one assign a fossil bone to a distinct place in human history? Apes vary widely, too; australopithecines may simply be a type that became extinct. As science journalist Roger Lewin notes:

> It is an unfortunate truth that fossils do not emerge from the ground with labels already attached to them. And it is bad enough that much of the labeling was done in the name of egoism and a naive lack of appreciation of variation between individuals; each nuance in shape was taken to indicate a difference in type rather than natural variation within a population.[27]

Likewise, if evolutionists find "simple" tools at a fossil site, they associate them with older ancestors—but today, people can still be found using simple stone tools (sometimes they're the best tool for a job), and it doesn't mean the users are "less evolved."

Sherlock Holmes used to pick up an object and say something like, "Watson, I deduce that the owner of this cigar was an Italian, six feet two inches tall, between forty and fifty years old, who walked with a limp and had a large mole on his left cheek." The speculative pronouncements of paleoanthropologists about bones and tools remind me somewhat of Holmes—but at least the Baker Street sage was describing someone alive, by whom he could verify his conclusions. Science primarily addresses the present. It cannot observe the past with the same authority. Dr. Greg Kirby, senior lecturer in population biology at Flinders University in Australia, commented:

> [N]ot being a paleontologist, I don't want to pour too much scorn on paleontologists, but if you were to spend your life picking up bones and finding little fragments of head and little fragments of jaw, there's a very strong desire there to exaggerate the importance of those fragments. . . .[28]

In 1983, *New Scientist* reported how a bone, thought to be the collarbone of an ancient hominid, actually turned out to be part of a dolphin's rib. The article quoted Dr. Tim White, anthropologist at the University of California, Berkeley: "The problem with a lot of anthropologists is that they want so much to find a hominid that any scrap of bone becomes a hominid bone."[29] As creationist Marvin Lubenow notes, "No one will care if you discover the oldest fossil broccoli, but if you are fortunate enough to discover the oldest fossil human, the world will beat a path to your door."[30]

Paleoanthropology is a field marked by jealousy and competition. Harvard's Hooton wrote: "The tendency towards aggrandizement of a rare or unique specimen on the part of its finder or the person to whom its initial scientific description has been entrusted, springs naturally from human egoism and is almost ineradicable."[31] We have already mentioned the well-known vanity of Eugene Dubois. In the nineteenth century, noted fossil hunters Othniel Marsh and Edward Drinker Cope engaged in a notorious personal war, each trying to outdo the other, with public charges and counter-charges of plagiarism and incompetence. Lewin recounts:

> So keen was the desire of each to score a point against the other that they would hire people to buy fossils from the sources of their competitor. And such was the race to name new species before the other did that they tried telegraphing the appropriate messages back East, often with the most hilarious garbled result.[32]

More recently, disagreements have occurred between "skull 1470" discoverer Richard Leakey and "Lucy" discoverer Donald Johanson about where their bones fit in man's history. In 1983, *The Weekend Australian*, referring to comments by Richard Leakey, noted:

> Echoing the criticism made of his father's *habilis* skulls, he added that Lucy's skull was so incomplete that most of it was "imagination made of plaster of paris," thus making it impossible to draw any firm conclusion about what species she belonged to.[33]

A 1995 *Wall Street Journal* article told much about paleoanthropology's state. It noted that Johanson had taken a starring role on the *Nova* TV series "In Search of Human Origins," which

> seemed the perfect vehicle to propel the handsome paleoanthropologist to the celebrity-scientist ranks of Carl Sagan. But others at the institute [of Human Origins] were uncomfortable with Dr. Johanson's aspirations. . . .
>
> No one objected to Dr. Johanson's "Nova" project more than 75-year old Garniss Curtis. Dr. Curtis shepherded the institute's team of eight geochronologists, the scientists who determine the age of fossils. Professional jealousies had long divided the two scientific camps, in part because the guys with the shovels and bones tend to hog the limelight. . . .
>
> A few weeks later, while munching on soft-shell crab at Berkeley's ritzy Chez Panisse restaurant, Dr. Johanson spotted Dr. Curtis and another of the institute's geochronologists wining and dining an elderly San Francisco socialite whom Dr. Johanson was trying to cultivate as a donor. An angry Dr. Johanson snubbed the group, much to the chagrin of members of the prestigious Leakey Foundation who were also at the table with Dr.

> Curtis. "Frosty glances all around" is how one person described the encounter. Later, Dr. Johanson lit into Dr. Curtis. . . .
>
> In September, the geochronologists dealt Dr. Johanson a tough professional blow: They published a paper showing that an Ethiopian skeleton found by Mrs. Getty's mentor, Dr. White, is 4.4 million years old. "Johanson is insanely jealous that his Lucy is no longer the oldest human ancestor," Dr. Renne says.
>
> And what of Lucy? Fingering the casts of her ribs in a wooden drawer at the institute, one bespectacled scientist observes: "Anthropologists dig up these bones so they can beat each other over the head with them."[34]

J. S. Jones wrote in *Nature* in 1990:

> Paleoanthropologists seem to make up for a lack of fossils with an excess of fury, and this must now be the only science in which it is still possible to become famous just by having an opinion. As one cynic says, in human paleontology the consensus depends on who shouts loudest.[35]

Over the years, there have been other proposed fossil men we have not had time to discuss, such as *Ramapithecus*, Heidelberg Man and Rhodesian Man. The human evolutionary scenario now runs something like this: *Australopithecus* (of which several varieties are described) eventually turned into *Homo habilis*, who became *Homo erectus*, which presumably included Java Man and Peking Man, and he in turn evolved into Neanderthals, then Cro-Magnon Man, and finally modern man.

But doubt and disagreement mark all of this. As Christopher B. Stringer noted in *Scientific American* in 1993: "The study of human origins seems to be a field in which each discovery raises the debate to a more sophisticated level of uncertainty."[36] Bones not fitting expected patterns have caused confusion, renaming of species, and creation of new ones. As we have illustrated, modern-looking bones often date older than the primitive.

In 1996 the *New York Times* reported new research determining that *Homo erectus* lived in Java as recently as 27,000 years ago. "This surviving population of H. erectus in Indonesia," the *Times* said, "would have been alive at the same time as anatomically modern humans—Homo sapiens—and also Neanderthals, whose exact place in human evolution is the subject of endless debate."[37] The *Times* noted that other scientists sharply contested the finding. Elsewhere, *Homo erectus* fossils are found contemporary with the ancient *Homo habilis* ("2 million years old"). Bones that morphologically fit *Homo erectus* have been dated a million years apart, yet look indistinguishable, which is hard for Darwinism: How could no evolution occur over such a long period?

Most textbooks avoid showing *comprehensive* tables of the discovered human fossils—doing so exposes the contradictions. Marvin Lubenow's book *Bones of Contention* exhibits such charts, documenting the innumerable disagreements and ambiguities in the fossil record.

Human evolution is not an orderly picture because it is highly subjective. As one wag put it, the only evolution that's seemingly taken place has been in the theories themselves. Dr. Robert Martin, a senior research fellow at the Zoological Society of London, stated in 1977: "In recent years several authors have written popular books on human origins which were based more on fantasy and subjectivity than on fact and objectivity."[38] Lowenstein and Zihlman later noted in *New Scientist*:

> The subjective element in this approach to building evolutionary trees, which many palaeontologists advocate with almost religious fervor, is demonstrated by the outcome: there is no single family tree on which they agree. On the contrary, almost every conceivable combination and permutation of living and extinct hominoids had been proposed by one cladist or another.[39]

Science News commented:

> If placed on top of one another, all these competing versions of our evolutionary highways would make the Los Angeles freeway system look like County Road 41 in Elkhart, Indiana.[40]

And just how much evidence exists, in total, forming the basis of these highways? The answer may surprise you. Lyall Watson wrote in *Science Digest*:

> The fossils that decorate our family tree are so scarce that there are still more scientists than specimens. The remarkable fact is that all the physical evidence we have for human evolution can still be placed, with room to spare, inside a single coffin![41]

John Reader used slightly different imagery in *New Scientist*:

> The entire hominid collection known today would barely cover a billiard table, but it has spawned a science because it is distinguished by two factors which inflate its apparent relevance far beyond its merits. First, the fossils hint at the ancestry of a supremely self-important animal—ourselves. Secondly, the collection is so tantalisingly incomplete, and the specimens themselves often so fragmentary and inconclusive, that more can be said about what is missing than about what is present.[42]

Arranging a few bone fragments in a sequence does not prove a relationship. Even if you had the complete fossilized bones of a father and son, you could not prove they were linked except by historical records. So it seems a bit pretentious to take a three-and-a-half-foot tall apelike creature, supposedly three million years old, and declare with certainty that we are its descendants.

Man is physically quite different from apes. Biology professor John W. Klotz noted:

> The human nose has a prominent bridge and an elongated tip which is lacking in the apes. . . . Man has red lips formed by an outrolling of the mucous membrane which lines the inside of his mouth; apes do not have this. Apes have thumbs on their feet as well as on their hands. . . . Man has the greatest weight at birth in relation to his weight as an adult. Yet at birth he shows the least degree of maturation and is by far the most helpless of creatures. Man's head is balanced on top of his spinal column; the head of the ape is hinged at the front instead of on top.[43]

And as creationist John Whitcomb observes, "while the physical differences between men and primates are quite great, *the spiritual/mental/linguistic/cultural differences are little short of infinite*."[44]

Noam Chomsky of MIT, acknowledged as the world's foremost linguist, remarked that "human language appears to be a unique phenomenon, without significant analogue in the animal world."[45] What animal has a complex vocabulary? Uses written symbols? Observes laws of grammar? Thinks up rhymes?

What animal manufactures tools, practices hobbies, composes music, enjoys artwork, solves math equations, laughs at jokes, thinks abstractly, obeys a conscience, or worships God? How can a being so distinct as man be written off as "just one more animal"?

In most sciences, a theory results from a series of observations. In the case of man's common ancestry with apes, the theory *preceded* the observations. Darwinists began with an assumption, then went searching for proof. The missing links are still missing. Why? For a profound reason. They were never there.

PLATE 29. Early artist's conception of Taung Child contrived to combine ape and human features.

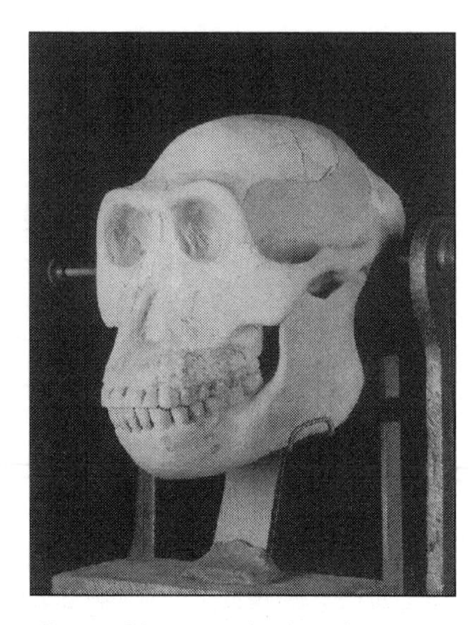

PLATE 30. Australopithecine skull

PLATE 31. British anatomist Solly Zuckerman found no links between man and australopithecines.

Plates 32 & 33. Jealousy between fossil hunters Othniel Marsh (left) and Edward Drinker Cope (right) spilled over into a media war. Cope was behind a *New York Herald* article that quoted one source as saying Marsh "has never been known to tell the truth when a falsehood would serve the purpose as well." Marsh counterattacked with an article claiming that Cope had assembled the skeleton of an extinct reptile rear-end first.

Plate 34. Richard Leakey with "skull 1470"

CHAPTER 10

Old Myths Never Die—They Only Fade Away

Before closing our look at Darwin's theory, we need to "mop up" some lingering falsehoods it has spawned.

The law that wasn't

When I went to school in the sixties, our biology textbooks showed a picture of a human embryo next to various animal embryos. The human looked almost indistinguishable from the animals. We were told this demonstrated the common ancestry we share with them.

It was further stated that embryonic development proved Darwinism, because the embryo went through various stages mimicking its evolutionary history. The fetus began as a single cell—just as life had billions of years ago. It then would undergo a tadpole stage, fish stage, amphibian stage, and so forth, en route to becoming human.

This theory was known as "embryonic recapitulation" as well as "the biogenetic law." Although the concept preceded Darwin (he discussed it in *The Origin of Species*), it was popularized by Ernst Haeckel. Haeckel, as we have mentioned, published baseless drawings of ape-men and "spontaneously generated" bacteria. He also created those famous pictures of identical-looking human and animal embryos. Haeckel explained:

> When we see that, at a certain stage, the embryos of man and the ape, the dog and the rabbit, the pig and the sheep, though recognizable as higher vertebrates, cannot be distinguished from each other, the fact can only be elucidated by assuming a common parentage. . . . I have illustrated this significant fact by a juxtaposition of corresponding stages in the development of a number of different vertebrates in my *Natural History of Creation* and in my *Anthropogeny*.[1]

But shamefully, Haeckel grossly altered the appearance of embryos to make his case. As Francis Hitching explained:

> But as a matter of biological fact, the embryos of men, apes, dogs, and rabbits are not at all the same, and can easily be distinguished by any competent embryologist. They only *looked* the same, in Haeckel's books, because he had chopped off bits here and there, and added bits elsewhere, to make them seem identical.
>
> Another example was his illustration of the "worm-like" stage through which all vertebrates were supposed to have passed. He published three identical drawings captioned respectively a dog, a chicken, and a tortoise. In 1886, a Swiss professor of zoology and comparative anatomy complained that Haeckel had simply used the same woodcut (of a dog embryo) three times.
>
> Over the years various other forgeries were exposed. To illustrate the "embryo of a Gibbon in the fish-stage," Haeckel used the embryo of a different kind of monkey altogether, and then sliced off those parts of the anatomy inconvenient to his theory, such as arms, legs, heart, navel and other non-fishy appendages.[2]

At Jena, the university where he taught, Haeckel was charged with fraud by five professors and convicted by a university court.[3] His deceit was thoroughly exposed in *Haeckel's Frauds and Forgeries* (1915), a book by J. Assmuth and Ernest J. Hull. They quoted nineteen leading authorities of the day. F. Keibel, professor of anatomy at Freiburg University, said that "it clearly appears that Haeckel has in many cases freely invented embryos, or reproduced the illustrations given by others in a substantially changed form."[4] L. Rütimeyer, professor of zoology and comparative anatomy at Basle University, called his distorted drawings "a sin against scientific truthfulness deeply compromising to the public credit of a scholar."[5]

Julius Weisner, professor of plant physiology at the University of Vienna, called Haeckel "one who in his most recent writings exhibits himself as a fanatical misleader of the people; one who, with delusive assurance, puts forth what have long been recognized as errors and mistakes as if they were verities."[6] J. Reinke, professor of botany at the University of Kiel, wrote that "wherever biology comes in, Haeckel uncritically jumbles together proved and unproved matter, and thus creates a chaos in the mind of his readers. It is the opinion of not a few that, on account of his lack of critical disposition, Haeckel forfeits all place in the ranks of serious naturalists."[7]

Such exposure did not prevent Haeckel's "biogenetic law" and fraudulent drawings from being spread in biology classrooms throughout the world. For decades to come, students were taught that the human embryo manifested reminders of man's past, such as "gill slits" from the fish stage of evolution. Actually, the "gill slits" evolutionists thought they saw were simply clefts and pouches which, as the embryo grows, develop principally into structures of the ear, jaw and neck.[8]

Many scientists knew Haeckel's theory was completely false. The human fetus is fully human at every stage. Keith Thomson, president of the Academy of Natural Sciences, wrote in *American Scientist*: "Surely the biogenetic law is as dead as a doornail. . . . As a topic of serious theoretical inquiry it was extinct in the twenties."[9] Dr. Sabine Schwabenthan wrote:

> Fetoscopy makes it possible to observe directly the unborn child through a tiny telescope inserted through the uterine wall. . . . The development of the child—from the union of the partners' cells to birth—has been studied exhaustively. As a result, long-held beliefs have been put to rest. We now know, for instance, that man, in his prenatal stages, does not go through the complete evolution of life—from a primitive single cell to a fish-like water creature to man. Today it is known that every step in the fetal developmental process is specifically human.[10]

Michael Richardson, an embryologist at St. George's Medical School, London, found there was no record that anyone *ever actually checked Haeckel's claims by systematically comparing human and other fetuses during development.* He assembled a scientific team that did just that—photographing the growing embryos of 39 different species. In a 1997 interview in London's *The Times,* Dr. Richardson stated:

> This is one of the worst cases of scientific fraud. It's shocking to find that somebody one thought was a great scientist was deliberately misleading. It makes me angry. . . . What he [Haeckel] did was to take a human embryo and copy it, pretending that the salamander and the pig and all the others looked the same at the same stage of development. They don't. . . . These are fakes. In the paper we call them "misleading and inaccurate," but that is just polite scientific language.[11]

Unfortunately, embryonic recapitulation's disproof is still not popularly known, and Haeckel's drawings continue to hold sway in the public mind. What Columbia University biologist Walter J. Bock noted in 1969 still seems true today: "[T]he biogenetic law has become so deeply rooted in biological thought that it cannot be weeded out in spite of its having been demonstrated to be wrong by numerous subsequent scholars."[12]

It seems the height of pretense that a theory—a fraudulent one, at that—was designated a "law," as if it had been established with the certainty of gravity. But this is symptomatic of Darwinism, where speculative opinions routinely masquerade as facts.

Based on this "law," it was claimed that babies born with growths on their tailbones had recapitulated tails—a throwback to our tree-swinging days. A. Rendle Short, professor of surgery at the University of Bristol, clarified this issue long ago:

> It is often stated that children are born with "tails"; but as a rule the alleged "tails" are nothing but fatty or fibrous tumors such as may be met with in many parts of the body, without any embryological significance. . . . There are many congenital abnormalities with which the medical profession is well acquainted: club foot, hare lip, cleft palate, congenital dislocations, naevi, supernumerary fingers and toes, spina bifida. But none of these recall the ape.[13]

The recapitulation theory greatly retarded understanding of fetal development. Children with Down's syndrome were called "mongoloids" because their development was thought to have arrested at the Mongoloid phase. (Older Darwinists thought human evolution had proceeded through Negroid, Mongoloid and Caucasoid stages. Gould notes: "Recapitulation provided a convenient focus for the pervasive racism of white scientists.")[14] It even inspired Sigmund Freud's idea that people "recapitulate" earlier evolutionary behavior. As Sir Gavin de Beer, professor of embryology at the University of London, aptly noted: "Seldom has an assertion like that of Haeckel's 'theory of recapitulation,' facile, tidy, and plausible, widely accepted without critical examination, done so much harm to science."[15]

Darwin's organ grinder

During the Scopes trial testimony, zoologist Horatio Hackett Newman, a defense witness, stated: "There are, according to Wiedersheim, no less than 180 vestigial structures in the human body, sufficient to make of a man a veritable walking museum of antiquities."[16] This was another of Darwinism's great myths: that the human body is loaded with vestigial organs—relics of the past no longer serving any purpose.

One reason why so many tonsillectomies were previously performed was the false belief that tonsils were "vestigial." Today it is recognized that the tonsils have an immune function. The thyroid gland, pituitary gland, thymus, pineal gland, and coccyx, also once considered useless, are now known to have important functions. The list of 180 "vestigial" structures is practically down to zero. Unfortunately, earlier Darwinists assumed that if they were ignorant of an organ's function, then it *had* no function.

Darwin had said that organs evolve over eons; a structure might therefore be incipient—on its way to becoming full-fledged. The "vestigial" idea thus posed quite a dilemma for evolution: Was a functionless organ "incipient" and on its way in, or "vestigial" and on its way out? The discovery that our organs *are* fully functional resolved the predicament, and suggested that Darwinism itself should be classed as vestigial.

Evolution evolves

We would be remiss if we overlooked the current vogue that evolution can happen quite rapidly. Darwin said:

> As natural selection acts solely by accumulating slight, successive, favourable variations, it can produce no great or sudden modifications; it can act only by short and slow steps.[17]

However, a number of twentieth-century evolutionists became disturbed by the lack of evidence supporting this—where were the transitional fossils? How could complex structures like the eye have evolved step-by-step? Gradualism seemed impossible.

In the 1940s, a number of geneticists, led by Richard Goldschmidt in America, and O. H. Schindewolf in Europe, began supporting a new explanation for evolution. They noted that mutations in an embryo sometimes produced monstrous creatures. Normally, such were quite poorly adapted to the environment. But was it possible that one might occasionally be born that wasn't? Could evolution thus occur in leaps?

Goldschmidt wrote in his 1940 book *The Material Basis of Evolution*: "A monstrosity appearing in a single genetic step might permit the occupation of a new environmental niche and thus produce a new type in one step."[18] He called this a "hopeful monster." Schindewolf wrote: "The first bird hatched from a reptilian egg."[19] This easy explanation did away with the difficulties of how reptilian scales evolved into bird feathers.

However, the "hopeful monster" theory was too much for most evolutionists to swallow. Even if such an organism was born by chance, it couldn't produce a new species unless it mated with *another* such creature, also generated by chance. Goldschmidt and his colleagues were ridiculed, and the theory discarded. As geneticist Sewall Wright noted in *Evolution*:

> I have recorded more than 100,000 newborn guinea pigs and have seen many hundreds of monsters of diverse sorts, but none were remotely "hopeful," all having died shortly after birth if not earlier.[20]

But, the fossil record problem would not disappear, and in 1977 Stephen Jay Gould published an article entitled "The Return of Hopeful Monsters." Although he did not agree that a bird could hatch from a reptile's egg, he predicted that "during the next decade Goldschmidt will be largely vindicated in the world of evolutionary biology."[21]

In a piece co-authored with Niles Eldredge, Gould introduced an updated mechanism for rapid evolution called "punctuated equilibrium."[22] A number of evolutionists today accept this trendy theory. It states that a species persists unchanged—in stasis or "equilibrium"—for millions of years. But then, a small segment of the species becomes isolated. It rapidly evolves, then takes over the ecological niche of the parent species, which becomes extinct. Since only a few individuals were involved in the fast evolution process, the fossil record does not show their transitional forms. The new species itself then persists for a long time at equilibrium, until "punctuated" by the arrival of another new, swiftly evolved species.

"Punctuated equilibrium" has an intellectual ring, and presents an explanation for why the fossil record shows stasis and lacks transitions. But without much supporting evidence, it seems little more than a rationalization. Darwinian errors, such as the belief that mutations increase genetic information, apply no less to this theory than to gradualism. As biologist Robert E. Ricklefs noted in *Science*: "The punctuated equilibrium model has been widely accepted, not because it has a compelling theoretical basis but because it appears to resolve a dilemma."[23] It was not born from evidence for evolution—but lack of it.

While punctuated equilibrium explains away fossil gaps between species, it certainly cannot account for the missing links between larger classifications (genus, family, etc.). It does not explain, for example, the complete absence of transitional fossils between invertebrates and fishes—a span that supposedly took 100 million years. Nor does it illuminate how complex organs evolved.

Years ago, when confronted with the lack of evidence for their beliefs, Darwinists responded that it takes eons for natural selection to develop a new species—therefore we couldn't see evolution because it happens so slowly. Now Gould and company say we can't see evolution because it happens so *rapidly*.

Through a bias darkly

But perhaps the truth is, we cannot see evolution because, beyond minor variations within species, it doesn't exist. Some scientists, while continuing to accept it, have at least recognized the weaknesses of Darwinian theory. Kenneth Hsu wrote in the *Journal of Sedimentary Petrology* (1986):

> We have all heard of *The Origin of the Species*, although few of us have had time to read it; I did not secure a copy until two years ago. A casual perusal of the classic made me understand the rage of Paul Feyerabend. . . . I agree with him that Darwinism contains "wicked lies"; it is not a "natural law" formulated on the basis of factual evidence, but a dogma, reflecting the dominating social philosophy of the last century.[24]

Swedish biologist Søren Løvtrup said in 1987:

> I believe that one day the Darwinian myth will be ranked the greatest deceit in the history of science.[25]

Over the years, other scientists have renounced evolution—or were never fooled to begin with. Sir John William Dawson, who pioneered Canadian geology and served as president of both McGill University and the British Association for the Advancement of Science, said:

> This evolutionist doctrine is itself one of the strangest phenomena of humanity. . . . a system destitute of any shadow of proof, and supported merely by vague analogies and figures of speech . . . . Now no one pretends that they rest on facts actually observed, for no one has ever observed the production of even one species. . . . Let the reader take up either of Darwin's great books, or Spencer's "Biology," and merely ask himself as he reads each paragraph, "What is assumed here and what is proved?" and he will find the whole fabric melt away like a vision. . . . We thus see that evolution as an hypothesis has no basis in experience or in scientific fact, and that its imagined series of transmutations has breaks which cannot be filled.[26]

Paul Lemoine, who was president of the Geological Society of France and director of the Natural History Museum in Paris, abandoned evolution. As chief editor of the *Encyclopedie Française*, 1937 edition, he wrote in that work:

> The theory of evolution is impossible. At base, in spite of appearances, no one any longer believes in it. . . . Evolution is a kind of dogma which the priests no longer believe, but which they maintain for their people.[27]

Dr. T. N. Tahmisian of the U.S. Atomic Energy Commission said in 1959:

> Scientists who go about teaching that evolution is a fact of life are great con-men, and the story they are telling may be the greatest hoax ever. In explaining evolution, we do not have one iota of fact.[28]

Zoologist Albert Fleischmann of the University of Erlangen declared: "The Darwinian theory of descent has not a single fact to confirm it in the realm of nature. It is not the result of scientific research, but purely the product of imagination."[29] He explained:

> [T]he theory suffers from grave defects, which are becoming more and more apparent as time advances. It can no longer square with practical scientific knowledge, nor does it suffice for our theoretical grasp of the facts. . . . No one can demonstrate that the limits of a species have ever been passed. These are the Rubicons which evolutionists cannot cross. . . . Darwin ransacked other spheres of practical research work for ideas. In particular, he borrowed his views on selection from T. R. Malthus' ideas regarding the dangers of overpopulation, to which he added the facts recorded by breeders. . . . But his whole resulting scheme remains, to this day, foreign to scientifically established zoology, since actual changes of species by such means are still unknown.[30]

Louis Bounoure, former director of the Strasbourg Zoological Museum, and later director of research at the French National Center of Scientific Research, stated in 1984:

> Evolutionism is a fairy tale for grown-ups. This theory has helped nothing in the progress of science. It is useless.[31]

Dr. Wolfgang Smith, who taught at MIT and UCLA, and has written on a wide spectrum of scientific topics, said in 1988:

> And the salient fact is this: *if by evolution we mean macroevolution* (as we henceforth shall), *then it can be said with the utmost rigor that the doctrine is totally bereft of scientific sanction.* Now, to be sure, given the multitude of extravagant claims about evolution promulgated by evolutionists with an air of scientific infallibility, this may indeed sound strange. And yet the fact remains that there exists to this day not a shred of *bona fide* scientific evidence in support of the thesis that macroevolutionary transformations have ever occurred.[32]

One of the most startling quotations comes from Colin Patterson, senior paleontologist at the British Museum of Natural History. He made the following comments during a keynote address at the American Museum of Natural History in 1981. Phillip Johnson, author of *Darwin on Trial*, elaborates: "Patterson came under heavy fire from Darwinists after somebody circulated a bootleg transcript of the lecture, and he eventually disavowed the whole business." Nonetheless, Johnson adds: "I discussed evolution with Patterson for several hours in London in 1988. He did not retract any of the specific skeptical statements he has made, but he did say he continues to accept 'evolution' as the only conceivable explanation for certain features of the natural world."[33] Here are Patterson's remarks:

> One of the reasons I started taking this anti-evolutionary view, or let's call it a non-evolutionary view, was last year I had a sudden realization for over twenty years I had thought I was working on evolution in some way. One morning I woke up and something had happened in the night, and it struck me that I had been working on this stuff for twenty years and there was not one thing I knew about it. That's quite a shock to learn that one can be misled so long. Either there was something wrong with me or there was something wrong with evolutionary theory. Naturally, I knew there was nothing wrong with me, so for the last few years I've tried putting a simple question to various people and groups of people.
>
> Question is: Can you tell me anything you know about evolution, any one thing, any one thing that is true? I tried that question on the geology staff at the Field Museum of Natural History and the only answer I got was silence. I tried it on the members of the Evolutionary Morphology seminar in the University of Chicago, a very prestigious body of evolutionists, and all I got there was silence for a long time and eventually one person said, "I do know one thing—that it ought not to be taught in high school."[34]

Adolf Hitler said people will believe a big lie more than a little one—a principle he used on his own followers. Evolution became a big lie. Facts do not always "speak for themselves"—rather, they are subject to interpretation. Once Darwinism was broadly accepted, many scientific fields began interpreting facts in an evolutionary context; this tended to validate the lie and make it even bigger. People have believed it, much as they believe a movie's events are taking place as they sit in a theater.

But it's time to turn off the projector. The lights are coming on for all the deceived generations. From senior citizens to teenagers, we're piling out of the cinema. Twenty-three skidoo, Darwin. Take a powder, Charley. Like, split the scene, daddio. *Origin of Species*?—total bummer, man.

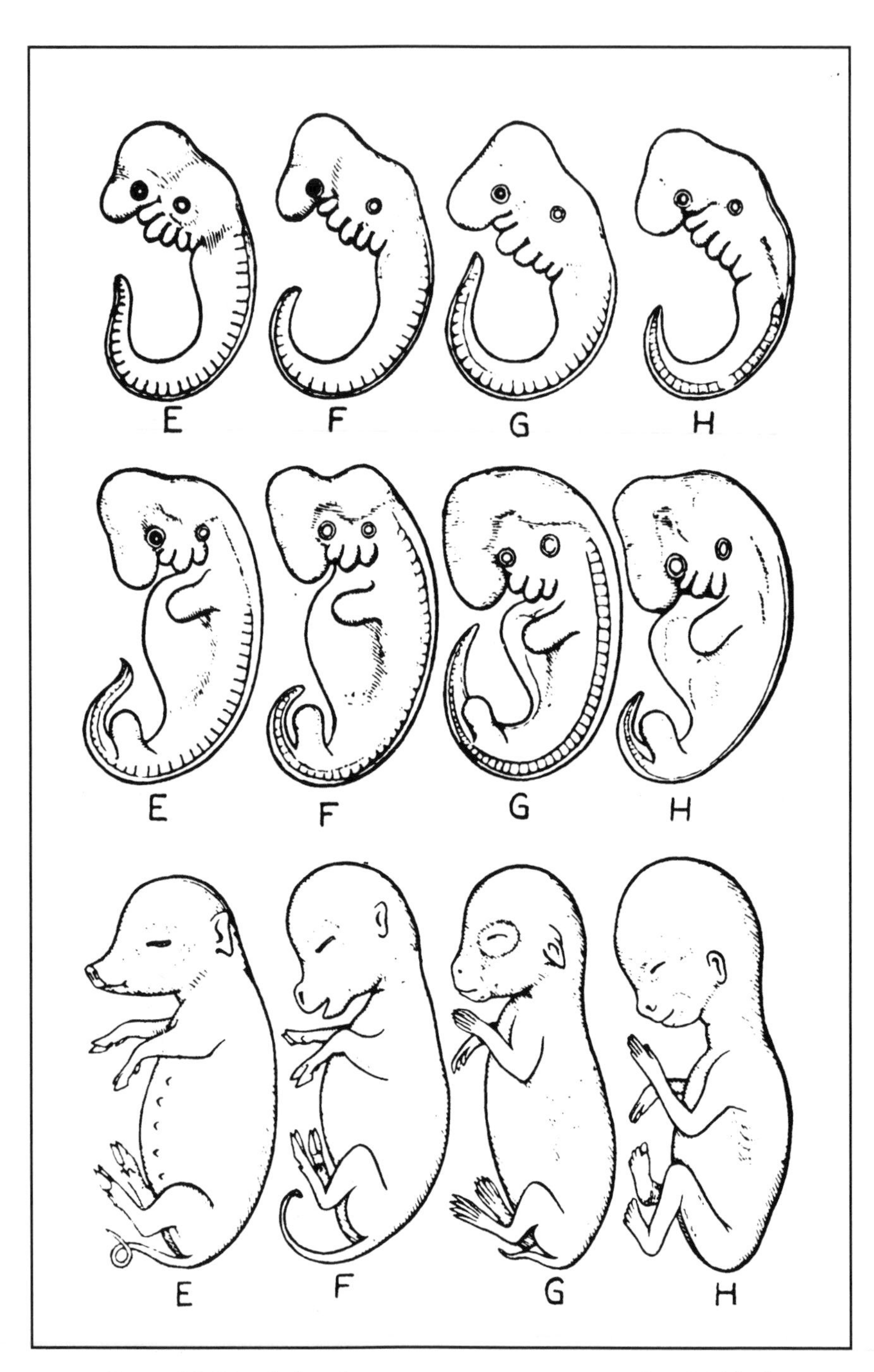

Plate 35. Haeckel embryo sequence, now exposed as falsified, purported to show (left to right) hog, calf, rabbit, human.

PLATE 36. Ernst Haeckel

PLATE 37. Stephen Jay Gould helped father the concept of punctuated equilibrium.

CHAPTER 11

The Big Bang Goes Blooey

The most widely accepted theory of the universe's origin says that, at one time, all mass and energy were compressed in a tiny "cosmic egg." Then, about fifteen billion years ago, the egg exploded, creating the universe in the Big Bang. Supposedly, the explosion first resulted in radiation, which later became matter.

Evidence for the Big Bang consists largely of (1) detection of cosmic radiation reaching the Earth uniformly from all directions, and (2) wavelength changes in the light emanating from stars ("red shift"), which suggests they are moving away from us and that the universe is therefore expanding.

The Big Bang has plenty going against it, however. Evolutionists have frequently stated that scientific discussions must exclude creation by God because supernatural events fall outside the realm of *natural law*, which is what science deals with.

But the Big Bang itself violates natural law. The laws of physics state that matter and energy can neither be created nor destroyed. This is the First Law of Thermodynamics, the law of conservation of energy. As the well-known physicist Paul Davies wrote in his book *The Edge of Infinity,* the Big Bang "represents the instantaneous suspension of physical laws, the sudden, abrupt flash of lawlessness that allowed something to come out of nothing. It represents a true miracle—transcending physical principles. . . ."[1]

If one allows for an event beyond natural law—a "true miracle" as Davies put it—then it is logically inconsistent to exclude other events, such as creation by God. If there was a "cosmic egg," who put it there? The cosmic chicken? Scientists have always agreed that there is a cause for every effect. How then can the greatest effect of all—the universe itself—have arisen without a cause?

Another reason evolutionists give for ignoring the Biblical version of creation is that it cannot, using scientific methods, be duplicated for observation and testing. But this same complaint may be made of the Big Bang.

One problem for the Big Bang is that explosions produce disorder. A hand grenade thrown into a building does not create anything regular, systematic, or harmonious. Yet these adjectives certainly apply to our universe. The Roman philosopher Cicero said more than 2,000 years ago:

> I cannot understand this regularity in the stars, this harmony of time and motion in their various orbits through all eternity, except as the expression of reason, mind and purpose. . . . Their constant and eternal motion, wonderful and mysterious in its regularity, declares the indwelling power of a divine intelligence. If any man cannot feel the power of God when he looks upon the stars, then I doubt whether he is capable of any feeling at all. . . . In the heavens there is nothing accidental, nothing arbitrary, nothing out of order, nothing erratic. Everywhere is order, truth, reason, constancy. . . .[2]

How could an explosion result in something as orderly as the solar system, with its orbiting planets? When we see it mimicked in the atom, whose electrons orbit a nucleus, we are reminded of the universe's harmony, even if the comparison is imperfect. In 1997, *Nature* reported a team of astronomers' findings about the organization of the cosmos:

> Startlingly, they see hints of a pattern: although the data are sketchy and the interpretation depends on details of the catalogue of clusters used, the network of superclusters and voids seems to form a three-dimensional lattice. . . .[3]

These "clusters and voids" present another dilemma for the Big Bang. Galaxy clusters are millions of light years apart. Why is there so much empty space between them? *Science 81* noted:

> As one sky scientist, IBM's Philip E. Seiden, put it, "The Standard Big Bang model does not give rise to lumpiness. That model assumes the universe started out as a globally smooth, homogeneous expanding gas. If you apply the laws of physics to this model, you get a universe that is uniform, a cosmic vastness of evenly distributed atoms with no organization of any kind." No galaxies, no stars, no planets, no nothin'. Needless to say, the night sky, dazzling in its lumps, clumps, and clusters, says otherwise.
>
> How then did the lumps get there? No one can say—at least not yet and perhaps not ever.[4]

Big Bang advocates say gravity caused gases to condense into the clumps we call galaxies, but this is only a hypothesis. Even the interpretations of "red shift" and cosmic radiation—the main evidences cited for the Big Bang—are disputed within the scientific world.

In his 1988 article in *Discover,* "The Big Bang Never Happened," Eric J. Lerner quoted Nobel Prize-winning physicist Hannes Alfvén on the Big Bang: "It is only a myth that attempts to say how the universe came into being. . . ."[5] As astronomer Fred Hoyle observed in *Science Digest*:

> As a result of all this, the main efforts of investigators have been in papering over holes in the big bang theory, to build up an idea that has become ever more complex and cumbersome. . . . I have little hesitation in saying that a sickly pall now hangs over the big bang theory. When a pattern of facts becomes set against a theory, experience shows that the theory rarely recovers.[6]

Perhaps God did create the cosmos in an expansive manner, furnishing some credence for the Big Bang, which at least agrees with the Bible on one point: that the universe had a beginning. Other explanations have been propounded, of course, such as Hoyle's own "steady state" theory, which conjectured that the universe has always existed as it does now. But this still fails to explain where matter came from.

Just your everyday, garden-variety, run-of-the-mill planet

According to evolutionists, Earth is no big deal. After all, look at the size of the cosmos. There are probably zillions of planets out there better than this one! We're just one more, orbiting the sun, which is just another star, in the Milky Way, which is just another galaxy. We're insignificant; this was much the point of Carl Sagan's book, *Pale Blue Dot.*

However, no astronomer has ever unequivocally seen a planet outside of our solar system. Distant stars having planets like our own is only an assumption, supported by limited and debatable evidence, such as variations in a star's light signal, interpreted to represent the gravity effects of a nearby planet. Writing in *Science* in 1998, Paul Kalas of the Max-Planck Institute for Astronomy decried the recent phenomenon of "'planet mania,' a bias among astronomers in which every cavity and blob, even a wiggle, in circumstellar dust disks is taken as evidence for extrasolar planets."[7]

Also, Earth is not so ordinary. Environmentalists frequently warn about the consequences of temperature changes. A couple more degrees of global warming, we are told, and lands will become parched, oceans will rise, catastrophic weather and ecological changes will occur. On the other hand, if temperatures cool a bit, here comes another ice age. Although some regard such warnings as hot air, they highlight the delicate balance permitting life on Earth.

Isaac Newton said: "Atheism is so senseless. When I look at the solar system, I see the earth at the right distance from the sun to receive the proper amounts of heat and light. This did not happen by chance."[8] If the Earth was closer to the sun, or farther away, we would either boil or freeze, and those environmental doom scenarios would materialize. Likewise, if the moon were much closer—or bigger—it would pull tides over the shorelines, destroying harbors and flooding plains.

But this is more than a distance matter. Earth, of course, rotates every 24 hours. That not only gives us night for rest and day for work, it keeps the planet evenly heated. Venus, though similar in size, rotates only once every 243 days. If we did the same, half our planet would scorch and the rest freeze.

The tilt of Earth's axis of rotation gives us seasons; the northern hemisphere slants toward the sun in summer, away from it in winter. Without this tilt, temperatures in any region would be the same in July as January. The planet would be less habitable, and farmland less productive.

The oceans absorb heat from the sun during the day, and release it at night, further stabilizing temperatures.[9] And anyone familiar with the greenhouse effect knows that Earth's atmosphere is a major temperature regulator.

The atmosphere also filters out ultraviolet light which would be deadly to life. And it burns up the meteors that approach Earth every day—with a thinner atmosphere, they would all strike the planet.

Oxygen makes up about 21 percent of our air. Without enough, we couldn't breathe, and would have no protective ozone layer. But with too much oxygen, the excess would burn up many biochemicals, adversely affecting life; fires would be abundant, and metals would quickly rust.

Only 0.03 percent of the atmosphere is carbon dioxide, but it's vital to plant growth and climate stability. Dramatically changing the concentration in either direction would be disastrous.

Then there is our water supply—vital to life, absent on other planets. Earth has a nice balance between oceans and dry land, but only because of our rugged topography. Without raised continents and low ocean basins—that is, if all terrain had a uniform elevation—then water would cover the Earth to a height of over one mile.[10]

What about gravity? Without it, there would be no tides and rain wouldn't fall. The atmospheric gases would be lost, we and everything else would fly off the planet, and the Earth would not revolve around the sun. Where did gravity come from? Natural selection acting on mutations?

Also, if Earth traveled much faster, centrifugal force would drive us away from the sun; much slower, we'd be drawn toward it.

Then there's our magnetic field, which protects us from cosmic rays and solar winds.

And there's the remarkable balance of living things. For example, if birds did not eat insects, keeping their population in check, insects would proliferate, and decimate the world's vegetation. (But birds supposedly didn't evolve until many millions of years after insects!)

We could go on, but clearly, innumerable circumstances must be just right for Earth to sustain life. The evolutionist has his stock explanation, of course: happy chance. Anyone want to figure the odds on all these conditions occurring together? Creation scientist Stuart E. Nevins put it well: "It is akin to supposing that Mona Lisa came into existence from globs of paint hurled at a canvas."[11]

Stellar dust goes bust

Evolutionary schemes for the *origin* of Earth, and the rest of the solar system, also face difficulties. An idea long popular was that the planets spun off from the sun. But why then does Earth rotate in the same direction as its orbit, while Venus rotates backwards? More than 98 percent of the sun is hydrogen and helium—the two lightest elements. If we came from the sun, why do we have such an abundance of heavier elements? As Fred Hoyle noted: "[W]e see that material torn from the sun would not be at all suitable for the formation of the planets as we know them. Its composition would be hopelessly wrong."[12]

Likewise, it is proposed that the various moons in our solar system spun off from their mother planets. But of the sixty-plus known satellites, more than twenty orbit *backwards* compared to their planets' orbit of the sun, while the rest move in the same direction. Why the discrepancies? If the moon came from Earth, why is it so different? As *Science Digest* noted after the Apollo lunar landings:

> To the surprise of scientists, the chemical makeup of the moon rocks is distinctly different from that of rocks on Earth. This difference implies that the moon formed under different conditions, Prof. [A. G. W.] Cameron explains, and means that any theory on the origin of the planets now will have to create the moon and the earth in different ways.[13]

Another hypothesis: the moon arrived from somewhere else, and Earth's gravity captured it. But why would an object that big, hurtling through space, simply stop and begin orbiting us? Why not rip past—or into—the Earth?

It is not just the Earth and moon that vary, but the entire solar system. As Richard A. Kerr observed in *Science* in 1994:

> The solar system used to be a simple place. Before any spacecraft ventured forth from the Earth, Venus seemed likely to be a warmer, wetter version of Earth. Small, more-distant Mars seemed chillier and drier, though conceivably habitable. Little Mercury might resemble Earth's moon, only hotter. And the four giant planets—all big balls of gas—presumably were much alike, except that those farther from the sun had less energetic weather.
>
> But 30 years of planetary exploration have replaced that simple picture with a far more complex image. "The most striking outcome of planetary exploration is the diversity of the planets," says planetary physicist David Stevenson of the California Institute of Technology. Ross Taylor of the Australian National University agrees: "If you look at all the planets and the 60 or so satellites, it's very hard to find two that are the same."[14]

The prevailing idea in recent years has been that a great cloud of interstellar gas and dust condensed into the sun and planets. This "nebular hypothesis" dates to the eighteenth century; temporarily out of favor, it is popular again. But it still does not explain the disparities in orbit and composition. And why did the dust and gas become several planets? Why not just condense into one great mass? Also, there's plenty of interstellar dust in our solar system now—why isn't *it* condensing? In fact, no one has ever seen a star or planet thus formed.

Astronomers Charles J. Lada and Frank H. Shu note in *Science* that "despite numerous efforts, we have yet to directly observe the process of stellar formation." "The origin of stars," they comment, "represents one of the most fundamental unsolved problems of contemporary astrophysics."[15] Abraham Loeb of Harvard's Center for Astrophysics states: "The truth is that we don't understand star formation at a fundamental level."[16] Professor Rogier A. Windhorst of Arizona State University says, "Nobody really understands how star formation proceeds; it's really remarkable."[17]

It is easy to invent a hypothesis to explain something. As we have seen, Darwinists have done this with evolution, concocting one scenario after another, with little or no evidence, to explain how natural selection created an organ. Unfortunately, cosmological theories, like those of natural selection, too often become "fact" once enunciated.

To us, stars are pinpoints of light; there is just so much we can learn about them. Sir Arthur Stanley Eddington, the astronomer who pioneered the study of stars' internal structure, observed:

> For the reader resolved to eschew theory and admit only definite observational facts, *all* astronomical books are banned. *There are no purely observational facts about the heavenly bodies.* Astronomical measurements are, without exception, measurements of phenomena occurring in a terrestrial observatory or station; it is only by theory that they are translated into knowledge of a universe outside.[18]

Perhaps the innumerable conditions permitting life on Earth are there because God created them. Perhaps the sun is there to give us light and heat; the moon to regulate the tides (and, just as the Bible says, to provide us with a "lesser light to govern the night").[19]

Design is suggested not only by the Earth, but by the very structure of the universe. As *Newsweek* noted:

> Physicists have stumbled on signs that the cosmos is custom-built for life and consciousness. It turns out that if the constants of nature—unchanging numbers like the strength of gravity, the charge of an electron and the mass of a proton—were the tiniest bit different, then atoms would not hold together, stars would not burn and life would never have made an appearance.[20]

The eminent British astronomer James Jeans wrote:

> [F]rom the intrinsic evidence of his creation, the Great Architect of the Universe now appears as a pure mathematician. . . . We discover that the universe shows evidence of a designing or controlling power. . . .[21]

Rocket scientist Wernher von Braun, who became director of NASA's space flight center and masterminded many of our first ventures into space and to the moon, declared:

> Manned space flight is an amazing achievement, but it has opened for mankind thus far only a tiny door for viewing the awesome reaches of space. An outlook through this peephole at the vast mysteries of the universe should only confirm our belief in the certainty of its Creator. I find it as difficult to understand a scientist who does not acknowledge the presence of a superior rationality behind the existence of the universe as it is to comprehend a theologian who would deny the advances of science.[22]

PLATE 38. Designed for living

PLATE 39. Star Spec

CHAPTER 12

Earth, Dahling, You Don't Look a Day Over Five Billion

Nah, only four and a half billion, they now say. Oh, yeah?

Here comes the sun

The sun is aging; it is not a constant. As it gives us energy, it *loses* that energy, at four million tons per second.[1] *Physics Today* reported in 1979:

> By analyzing data from Greenwich [Observatory] in the period 1836–1953, John A. Eddy (Harvard-Smithsonian Center for Astrophysics and High Altitude Observatory in Boulder) and Aram A. Boornazian (a mathematician with S. Ross and Co. in Boston) have found evidence that the Sun has been contracting about 0.1% per century during that time, corresponding to a shrinkage rate of about 5 feet per hour. And digging deep into historical records, Eddy has found 400-year-old eclipse observations that are consistent with such a shrinkage.[2]

These were stunning numbers, for extrapolation meant that 100,000 years ago, the sun would have been about twice its size, making life basically untenable. And 100 million years ago, when dinosaurs were supposedly roaming about, the sun's perimeter would have extended well beyond where Earth is. Eddy later reduced his estimate of shrinkage from five to two feet per hour—but that's still far too much for evolution. This disclosure alone, if correct, shoots down both the alleged age of our planet and Darwin's theory. Evolutionists reason that the sun's shrinkage must be temporary, cyclical, or peripheral, or that the observations were erroneous.

Evolutionists get mooned

The moon's recession from Earth also poses a time difficulty for evolution. As Dr. Jonathan Sarfati notes:

> [T]he moon is slowly receding from Earth at about 4 cm (1 1/2 inches) per year, and the rate would have been greater in the past. The moon could never have been closer than 18,400 km (11,500 miles), known as the *Roche Limit*, because Earth's tidal forces (i.e., the result of different gravitational forces on different parts of the moon) would have shattered it. But even if the moon had started receding from being in contact with the earth, it would have taken only 1.37 billion years to reach its present distance.[3]

Comety of errors

Comets were presumably formed at the same time as the solar system. But they disintegrate considerably as they orbit near the sun, as evidenced by their long tails, some of which extend up to 100 million miles. Some comets have been observed breaking apart. How could they last for five billion years? Astronomer Fred Hoyle wrote: "It is an immediate inference that these comets cannot have been moving around the sun as they are at present for much longer than a million years, since otherwise they would already have broken up."[4]

In 1950, to resolve the problem, Dutch astronomer Jan Oort proposed that there exists a far-away cloud containing billions of comets. They supposedly lie dormant in this "storehouse" until some cosmological event dislodges one, sending it into our solar system. The imaginary storehouse has been named "the Oort cloud." The 1998 *World Book Encyclopedia* matter-of-factly states: "Long-period comets arrive from the *Oort cloud*, a collection of comets 1,000 times farther away than Pluto's orbit"—not telling readers that the Oort cloud is a speculation that no one has ever seen. Short-period comets (those that take less than 200 years to orbit the sun) are now said to originate in the "Kuiper belt," whose existence has also been debated.

Missing meteorites

By comparing the number of meteorites in the earth to the rate at which they fall, we should be able to roughly estimate our planet's age. Meteorites' scarcity suggests that Earth is only a few thousand years old. They are mostly found in very recent (uppermost) terrain. If the geologic layers slowly formed over millions of years, as evolutionists say, they should be full of meteorites. F. A. Paneth noted in *Vistas In Astronomy*:

> The quantity of coal mined during the last century amounted to many billions of tons, and with it about a thousand meteorites should have been dug out, if during the time the coal deposits were formed the meteorite frequency had been the same as it is today. Equally complete is the absence of meteorites in any other geologically old material that has been excavated in the course of technical operations.[5]

Back in 1932, W. A. Tarr of the University of Missouri wrote in *Science*:

> For many years, I have searched for meteorites or meteoric material in sedimentary rocks [which is what the geologic column chiefly consists of]. . . . I have interviewed the late Dr. G. P. Merrill, of the U.S. National Museum, and Dr. G. T. Prior, of the British Natural History Museum, both well-known students of meteorites, and neither man knew of a single occurrence of a meteorite in sedimentary rocks. . . . This letter is a petition for any information indicating that meteorites do occur in the sediments.[6]

The petition went unanswered. Few meteorites have been found below the highest levels of the geologic column.

It's a gas

Helium leaks through the Earth's crust at a measurable rate and enters the atmosphere, of which it comprises only a minute fraction—five parts per million. Some helium also escapes into outer space, but it must overcome gravity's pull, and cannot escape as quickly as it enters. The problem? There should be a lot more if Earth is billions of years old. The dilemma prompted an article in *Nature* entitled "Where is the Earth's Radiogenic Helium?"[7] and one in *New Scientist* entitled "What Happened to the Earth's Helium?" The latter noted: "At first sight there ought to be about a thousand times as much helium in the atmosphere as there is." After reviewing shortcomings in one proposed explanation, the article uneasily concluded:

> The only way we can explain this state of affairs, according to Dr. [E. E.] Ferguson [an atmospheric specialist], is to suppose that the atmosphere was somehow depleted of its helium a few million years ago and has been building up its concentration since. What form the catastrophic event might have taken he does not say. One possibility could be that it was a collision with a comet. . . .[8]

Dr. Larry Vardiman, chairman of the Physics Department at the Institute for Creation Research, doesn't buy evolution and therefore has no need to conjure up such rationalizations. His technical book, *The Age of the Earth's Atmosphere: A Study of the Helium Flux through the Atmosphere* makes a strong case that helium accumulation limits the Earth's age to about 10,000 years.[9]

Earth hits the skids

Earth's rate of rotation is slowly but measurably declining. Based on this, Lord Kelvin, the nineteenth-century physicist who introduced the Kelvin temperature scale, argued against an old planet. His point: if Earth were billions of years old, its centrifugal force would have been so great during its alleged molten state that the equatorial regions would have bulged out. This would give our planet a different shape, since it would have retained much of the bulging as it cooled and consolidated. Because the decline in rotational rate is now known to be greater than previously thought, Lord Kelvin's argument holds up even better.[10]

Drawing conclusions

The Earth is a giant magnet; that's why compasses work. In 1967, the U.S. Commerce Department published a detailed scientific study of the magnetic field, showing its intensity had decreased by seven percent since first being measured in 1829. The report predicted that if decay continued at the same rate, Earth's magnetic field would vanish by the year 3991.[11] *Science News* said of the field in 1980:

> Measurements of the main field, when plugged into a computer model, show that the over-all intensity of the field is declining at a rate of 26 nanoteslas per year. . . . If the rate of decline were to continue steadily—and [researcher Gilbert] Mead stressed that researchers cannot determine if it will— the field strength would reach zero in 1,200 years.[12]

And *Scientific American* reported in 1989:

> In the next two millennia, if the present rate of decay is sustained, the dipole component of the field should reach zero.[13]

This also means the field was previously stronger. If the planet is billions of years old, the magnetic field would have been impossibly high in ages past, making life unsustainable. Evolutionists point to archaeomagnetic data (measurements of magnetization in ancient artifacts) that indicate the field's intensity has fluctuated in the past. But Dr. Russell Humphreys, a physicist at Sandia National Laboratories, has proposed a model that shows how, even taking such fluctuations into account, the magnetic field may limit the Earth's age to roughly 10,000 years.[14]

Young Man River

The Mississippi River is said to have formed about two million years ago. However, a glance at a map of the USA shows the river has created a huge delta jutting into the Gulf of Mexico. In the nineteenth century, A. A. Humphreys and H. L. Abbot, of the U.S. Army Corps of Topographical Engineers, extensively studied the river and delta. They determined the latter's age by dividing its average annual growth into the distance from the continental shoreline. Their conclusion: "Adopting this rate of progress (262 feet per annum), four thousand four hundred years have elapsed since the river began to advance into the gulf."[15] In the olden days, the truth was simple to get at.

The people problem

One of the greatest obstacles to the Darwinian age of Earth: man himself. Supposedly, we've been evolving for millions of years. In recent decades, dark warnings have been issued about the dangers of overpopulation. Perhaps this idea's greatest exponent was Paul Ehrlich, author of *The Population Bomb.*

The famines Ehrlich predicted didn't materialize, but world population has undeniably been growing—from about one billion in 1800 to two billion in 1930 to some six billion today. Evolution's dilemma comes from looking further back and doing a little math. Creation scientist Henry Morris noted a few years ago:

> The average family size today, worldwide, is about 3.6 children, and the annual population growth is 2 percent. . . . assume an initial population of two people, the first parents. . . . The evolution model, with its million-year history of man, has to be strained to the breaking point. It is essentially incredible that there could have been 25,000 generations of men with a resulting population of only 3.5 billion. If the population increased at only 1/2 percent per year for a million years, or if the average family size were only 2.5 children per family for 25,000 generations, the number of people in the present generation would exceed 10^{2100}, a number which is, of course, utterly impossible . . . only 10^{130} electrons could be crammed into the entire known universe.[16]

Since we have been proliferating faster in recent years due to improved health care and nutrition, Morris did not use today's rates to estimate ancient population growth. He pointed out that world population was roughly 600 million in A.D. 1650 and one billion in 1800. This translates to annual growth of 1/3 percent. Assuming that as a norm, we would get the current population in 6,300 years if we started with just two people.[17]

The Bible, of course, reports a great Flood which left *eight* people in its wake, and theologians debate the Flood's precise date, but if we take the well-known one fixed by Archbishop James Ussher, 2350 B.C., we would get today's population from an annual growth rate of just 0.5 percent.

These calculations are simple and hypothetical. In reality, of course, eight people cannot increase by just 0.5 percent, and initial reproduction rates would have necessarily been much higher to account for the large civilizations that existed in ancient times. But the Biblical model adjusts to fall within mathematic reality far more easily than the long Darwinian one.

Evolutionists counter that the population must have waxed and waned over the eons, held in check by plagues, famines and warfare. But these have troubled man throughout recorded history, and have not stopped world population from growing. In fact, wars have been far more deadly in modern times, due to advances in weaponry. If people have been evolving for as long as Darwinists say, where did the bones go from those billions of bodies?

One clue to how long we've been around is history itself. Writing is believed to date back about 5,500 years. Before then, man was "prehistoric." But perhaps there is no record of those times because they *never were*. Man supposedly evolved for millions of years, finally reaching the point where he could write. At two, my son David knew the alphabet, which is not unusual. A six-year old can read and write; so how is it that 50,000 years ago, no adults were smart enough to devise a written language? Brain evolution sure must have happened suddenly.

And so forth

Why do we see an explosive gusher when a drill strikes oil? Because oil, like natural gas, is maintained in the earth at enormously high pressure—about 5,000 pounds per square inch at a depth of 10,000 feet. Supposedly oil and gas have been lying there for millions of years. But how could they have lasted that long without leaking or otherwise dissipating those extreme pressures?

As land undergoes erosion, sediment passes into the sea at a calculable rate. If the oceans were a billion years old, there should be about 100,000 feet of sediment blanketing the ocean floor.[18] Instead, there is less than 3,000 feet. Also, at the current pace of erosion, the continents would reduce to sea level in 14 million years.[19] Further havoc for evolution.

Comparing the ocean's salt content to its accumulation rate,[20] uranium content to accumulation rate,[21] and various other measures suggest a young Earth.

How do evolutionists justify their estimate of billions of years? Their greatest argument is probably Darwinian theory itself. Chance and natural selection would require eons to develop life forms. After one buys that argument, all evidence that doesn't fit is rejected. As British physicist H. S. Lipson observed in *Physics Bulletin*: "In fact, evolution became in a sense a scientific religion; almost all scientists have accepted it and many are prepared to 'bend' their observations to fit in with it."[22]

But what about radiometric techniques—carbon dating, uranium-lead, etc.? Don't they prove an ancient Earth? Because these tools tend to yield great ages—supporting evolution—they have received disproportionate attention compared to other valid methods of dating the planet. We shall devote our next chapter to them. What fun!

CHAPTER 13

Assumptions Aplenty

Carbon dating is perhaps the most famous radiometric dating technique. But it does not prove an old Earth. Even its advocates acknowledge it has an upper dating limit of about 50–60,000 years.

How does the method work? Carbon is found in all living things. Like many elements, it has more than one form or "isotope." Most is carbon-12 or C-12. The twelve refers to carbon's atomic weight. However, there is another, rare isotope—carbon-14, or C-14. It is created by cosmic rays interacting with nitrogen gas in the atmosphere. C-14 is also called "radiocarbon," and is the isotope crucial to the dating technique. There is about one C-14 atom for every trillion of the regular C-12 atoms. When plants absorb carbon dioxide from the air, they absorb C-14 along with it. Animals get C-14 by eating plants. Thus all living things contain C-14 in their tissues, and the concentration— ratio of C-14 to C-12— is the same as the atmospheric concentration.

When the plant or animal dies, however, the C-14 in its tissues starts decaying back into nitrogen, while the C-12 remains stable. It has been determined that after about 5700 years, half the C-14 atoms would be gone. C-14 thus has a "half-life" of 5700 years. After another 5700 years, a quarter of the atoms would be left, after another 5700, an eighth, and so on. Thus, by ascertaining the ratio of C-14 to C-12 in an object, scientists believe they can determine its age.

However, for anything approaching 100,000 years old, there would be no measurable carbon-14 left, so the technique cannot date to millions or billions of years. A further limitation: radiocarbon cannot date rocks—only tissues once living.

More importantly, carbon dating depends on various assumptions. Perhaps the most tenuous is that C-14's atmospheric concentration has always been constant. It is assumed that when organisms lived many thousands of years ago, the concentration would have been the same as today.

However, it is now generally recognized that atmospheric C-14 is rising. As Elizabeth K. Ralph and Henry M. Michael noted in their *American Scientist* article "Twenty-five Years of Radiocarbon Dating":

> We now know that the assumption that the biospheric inventory of C^{14} has remained constant over the past 50,000 years or so is not true.[1]

One reason may be Earth's declining magnetic field; this allows more cosmic radiation, which in turn creates more C-14. Objects containing little C-14 may appear "old" when in fact there was simply less atmospheric C-14 when they were living. Robert Stuckenrath, of the Smithsonian Institution's Radiation Biology Laboratory, observed in a piece entitled "Radiocarbon: Some Notes from Merlin's Diary":

> Now for one of my petter peeves. One of the basic assumptions of C-14 dating has been that the intensity of cosmic ray penetration, responsible for the production of C-14, has been stable for that period of time over which we likely can make radiocarbon measurements. Alas, and lackaday, 'tis not so. There have been variations in that intensity, and therefore variations in the C-14 content in the atmosphere.[2]

There are other factors. One is past temperature changes in the oceans—which absorb carbon dioxide and thus affect C-14 concentration. Another is fossil fuels (oil, coal, etc.), which are high in carbon and have been burned extensively since the industrial revolution, changing the atmospheric balance. Volcanoes, which spew carbon dioxide into the environment, and other past catastrophic events, may also have altered C-14's concentration.[3]

Of course, the greatest catastrophe the Bible describes is a worldwide Flood. If the atheists will permit us to indulge this scenario for a paragraph, Genesis states that when God made the Earth, there was great "water above." Some creation scientists interpret this as a canopy of water vapor surrounding the planet, later released at the time of the Flood. The geologic record indicates that lush vegetation once widely covered Earth (for example, palm tree fossils have been found in Alaska and fossils of tropical leaves just 250 miles from the South Pole).[4] This fits the theoretical water canopy, which would have provided a warm greenhouse effect (increasing vegetation and carbon dioxide) and shielded Earth from cosmic rays (decreasing C-14). Assuming the Flood did occur, little if any C-14 may have existed before then. This would give anything older than the Flood a false appearance of great age.

Another problem for carbon dating is that C-14 may have been washed out of material ("leached") or *added* through contamination. The tester must assume that the object has remained pristine for thousands of years. As Dr. Ernest Antevs noted in the *Journal of Geology*:

> In appraising C-14 dates, it is essential always to discriminate between the C-14 age and the actual age of the sample. The laboratory analysis determines only the amount of radioactive carbon present. . . . However, the laboratory analysis does not determine whether the radioactive carbon is all original or is in part secondary, intrusive, or whether the amount has been altered in still other irregular ways besides by natural decay. . . . the analyses, however numerous, tell us nothing about the vital ever present problem: Is the C-14 age also the actual age of the sample?[5]

Many absurd dates have been found. For example, in 1984 *Science* reported that the shells of living snails in artesian springs in Nevada were carbon-dated as 27,000 years old. This was "attributed to fixation of dissolved HCO_3 with which the shells are in carbon isotope equilibrium."[6]

In *Antarctic Journal*, geologist Wakefield Dort, Jr., related the effects of antarctic sea water, which has subnormal C-14 levels:

> [T]he apparent radiocarbon age of the Lake Bonney seal known to have been dead no more than a few weeks was determined to be 615 ± 100 years. A seal freshly killed at McMurdo had an apparent age of 1,300 years.[7]

Willard Libby, who invented the method, said in his article "Radiocarbon Dating" in *American Scientist*:

> The first shock Dr. Arnold and I had was that our advisors informed us that history extended back only 5000 years. We had thought initially that we would be able to get samples all along the curve back to 30,000 years, put the points in, and then our work would be finished. You read books and find statements that such and such a society or archaeological site is 20,000 years old. We learned rather abruptly that these numbers, these ancient ages, are not known; in fact, it is about the time of the first dynasty in Egypt that the last historical date of any real certainty has been established.[8]

Although the public is led to believe radiocarbon dating is nearly infallible, the method is controversial in scientific circles. J. Gordon Ogden, director of a radiocarbon laboratory, after enumerating a half-dozen factors that can falsify the process, noted in 1977:

> I find myself increasingly distressed that users of radiocarbon dates fail to understand or appreciate what the quoted figures really mean. . . . all that a date represents, is a "best estimate" of the radiocarbon content of the sample received by the laboratory. It includes none of the sampling or

> physical and biological errors sources mentioned earlier. . . . It may come as a shock to some, but fewer than 50 percent of the radiocarbon dates from geological and archaeological samples in northeastern North America have been adopted as "acceptable" by investigators.[9]

Robert E. Lee, assistant editor of the *Anthropological Journal of Canada,* summed the situation so thoroughly in that journal in 1981, that I think it best to quote him extensively:

> The troubles of the radiocarbon dating method are undeniably deep and serious. Despite 35 years of technological refinement and better understanding, the underlying *assumptions* have been strongly challenged, and warnings are out that radiocarbon may soon find itself in a crisis situation. Continuing use of the method depends on a "*fix-it-as-we-go*" approach, allowing for contamination here, fractionation there, and calibration whenever possible. It should be no surprise, then, that fully *half* of the dates are rejected. The wonder is, surely, that the remaining half came to be *accepted*. . . .
>
> Radiocarbon dating has somehow avoided collapse onto its own battered foundation, and now lurches onward with feigned consistency. The implications of pervasive contamination and ancient variations in carbon-14 levels are steadfastly ignored by those who base their argument upon the dates.
>
> The early authorities began the charade by stressing that they were "not aware of a single significant disagreement" on any sample that had been dated at different labs. Such enthusiasts continue to claim, incredible though it may seem, that "no gross discrepancies are apparent." Surely 15,000 years of difference on a single block of soil is indeed a *gross* discrepancy! And how could the excessive disagreement between the labs be called insignificant, when it has been the basis for the reappraisal of the standard error associated with each and every date in existence?
>
> Why do geologists and archaeologists still spend their scarce money on costly radiocarbon determinations? They do so because occasional dates *appear* to be useful. While the method cannot be counted on to give good, unequivocal results, the numbers do impress people, and save them the trouble of thinking excessively. . . .
>
> No matter how "useful" it is, though, the radiocarbon method is still not capable of yielding accurate and reliable results. There *are* gross discrepancies, the chronology is *uneven* and *relative*, and the accepted dates are actually *selected* dates.[10]

As Robert Stuckenrath summed it up: "After all, this whole blessed thing is nothing but 13th-century alchemy, and it all depends on which funny paper you read."[11]

Thus, radiocarbon is not only incapable of dating the Earth to billions of years, but shaky even for thousands. It is probably most reliable for very recent specimens. These are less likely to have had their C-14 content altered, more likely to have shared an atmosphere similar to ours, and easier to verify through historical records.

Other radiometric techniques

There are additional methods, similar in principle to carbon dating, that produce older dates evolutionists use to claim a great age for the Earth. They all involve unstable, radioactive isotopes, which decay into different elements over time. Again, isotopes are just forms of a chemical element. An isotope of uranium decays into an isotope of lead, an isotope of potassium decays into the gas argon, and an isotope of the element rubidium decays into strontium.

The beginning isotope, such as uranium, is called the "parent," and the end-product of decay, such as lead, the "daughter." Scientists believe that by comparing the amounts of parent to daughter in a given material, they can determine its age. The concept of radiometric dating is like an hourglass, with sand running from top to bottom. If there is minimal sand in the top of the hourglass (the "parent" element), and much sand in the bottom (the "daughter" element), an object is considered very old.

Like C-14, each radioactive isotope has a "half-life"—the time it takes for half the atoms to decay into the daughter. Uranium-238, which decays into lead, is believed to have a half-life of about 4.5 billion years, the approximate age assigned to Earth.

However, just as radiocarbon applies only to once-living matter, these other techniques have their own limitations. Seventy-five percent of the geologic column's rock is sedimentary (deposited chiefly by water, like limestone or sandstone). Radiometric methods do not work on sedimentary rock. They are effective only with "igneous" rock (that which was once hot and molten, like lava or deep granites) and "metamorphic" rock (that transformed by heat or pressure, such as marble). Since most fossils are lodged in sedimentary rock, they cannot be radiometrically dated. And for the relatively few fossils in igneous and metamorphic rock, even if the dating technique works, it doesn't necessarily reveal the fossil's age—only the age of the minerals, which may have formed long before the fossil! The bodies in a graveyard are not necessarily as old as the graveyard.

Like C-14, these methods depend on unsupported assumptions. One is the uniformity of decay rates. Have they always been constant? Half-lives are extrapolated from limited testing periods. Can we be certain that billions of years ago, uranium was decaying at the same rate? We have actually observed only about 0.000002 percent of its half-life.[12] Could past events have changed anything? Radiometric scientists try to use isotopes whose decay rates resist forces applied in laboratories. But as A. E. Wilder-Smith noted:

> Yet everyone today knows that the half-life of plutonium, for example, can be altered almost at will. The same applies to certain isotopes of uranium and other radioactive substances also. If the plutonium is in the environment of an atom bomb (i.e., exposed to strong neutron fluxes) the half-life of this radioactive metal may amount to a few nano-seconds instead of thousands of years. If the same metal is placed within an atomic reactor (i.e., subjected to a variable controllable neutron flux), then its half-life can be adjusted at will. If, due to cosmic events in the past, high neutron fluxes occurred on the earth's surface (e.g., before biogenesis), then it would have been possible for the half-lives of certain radioactive substances to be significantly altered.[13]

Geologist G. Brent Dalrymple is an authority on radiometric dating and a staunch defender of evolution. At the Arkansas creation-evolution trial of 1981, the ACLU called him as a witness to testify that the earth is billions of years old. The following cross-examination by Deputy Attorney General David Williams is revealing:

Q To the best of your knowledge, has the rate of radioactive decay always been constant?

A As far as we know from all the evidence we have, it has always been constant. We have no, either empirical or theoretical, reason to believe it is not.

Q So as far as you know, it would have been constant one billion years ago, the same as it is today.

A As far as we know.

Q Five billion years ago?

A As far as we know.

Q Ten billion years ago?

A As far as we know. . . . [This exchange continued for a while.]

Q Would the rate of radioactive decay have been constant at the time of the big bang?

A I am not an astrophysicist. I don't know the conditions that existed in the so-called primordial bowl of soup, and so I am afraid I can't answer your question.

Q So you don't have any opinion as to whether it was constant then?

A That's out of my field of expertise. I can't even tell you whether there were atoms in the same sense that we use that term now.[14]

As we noted in Chapter Eleven, the Big Bang, if it occurred, violated natural law, creating matter and energy from nothing. In keeping with this, Dalrymple acknowledged that when the universe evolved—fifteen billion years ago in Big Bang cosmology—we don't know if radioactive decay rates applied then.

So what if *God* created the universe, just as the Bible says? Like the Big Bang, this event would have transcended natural laws. Radioactive decay rates would not have applied then either. As Dalrymple would put it, we couldn't even know "whether there were atoms in the same sense that we use that term now." And these circumstances would have operated a few thousand years ago, not billions. Recent, dramatic events could have so impacted the cosmos that our assumptions about radioactive elements are falsified. As chemist Frederic B. Jueneman noted in *Industrial Research and Development*:

> The age of our globe is presently thought to be some 4.5 billion years, based on radiodecay rates of uranium and thorium. Such "confirmation" may be short-lived, as nature is not to be discovered quite so easily. There has been in recent years the horrible realization that radiodecay rates are not as constant as previously thought, nor are they immune to environmental influences.
>
> And this could mean that the atomic clocks are reset during some global disaster, and events which brought the Mesozoic to a close may not be 65 million years ago but, rather, within the age and memory of man.[15]

Other assumptions further weaken the longer dating methods. As with C-14, these elements risk leaching and contamination. How do we know water action has not washed out some of the parent or daughter elements over the purported eons? If there was a global flood, this probably would have occurred. Can we be certain argon gas did not leak from a specimen—

or that atmospheric argon did not contaminate it? Although radiometric scientists seek samples unlikely to have been compromised, rocks do not come with labels that say, "I have remained pure and undisturbed for three billion years."

How do we know that a specimen did not have daughter element to begin with? Did the rock's lead come from uranium—or was it primordial lead? Again, the analyst makes an assumption.

The unreliability of such methods is attested by the discrepant dates they yield. Evolutionist William Stansfield of California Polytechnic State University acknowledged:

> It is obvious that radiometric techniques may not be the absolute dating methods that they are claimed to be. Age estimates on a given geological stratum by different radiometric methods are often quite different (sometimes by hundreds of millions of years). There is no absolutely reliable long-term radiological "clock."[16]

Geologist Steven A. Austin dated samples from the Uinkaret Plateau, which contains some of the highest rocks in the Grand Canyon (making them positionally young). Using the rubidium-strontium method, he obtained an age of approximately 1.3 billion years. Lead dating has put the rocks at 2.6 billion years, and potassium-argon at anywhere from 10,000 to 117 million years. If the rubidium or lead methods are correct, these high rocks would be older than those lying near the canyon's base (the Cardenas Basalt, dated by different techniques from 0.7 to 1.1 billion years old).[17]

The *Journal of Geophysical Research* reported that lava from an 1800–1801 eruption of the Hualalai volcano in Hawaii had been radiometrically dated. Using various minerals and methods, ten different dates were derived, ranging from 140 million to 2.96 billion years.[18]

Sunset Crater, an Arizona volcano, is known from tree-ring dating to be about 1,000 years old. But potassium-argon put it at over 200,000 years.[19]

For the volcanic island of Rangitoto in New Zealand, potassium-argon dated the lava flows as 145,000 to 465,000 years old, but the journal of the Geochemical Society noted that "the radiocarbon, geological and botanical evidence unequivocally shows that it was active and was probably built during the last 1000 years." In fact, wood buried underneath its lava has been carbon-dated as less than 350 years old.[20]

Even the lava dome of Mount St. Helens has been radiometrically dated at 2.8 million years.[21] So one senses the futility of dating fossils by the rocks containing them.

After Richard Leakey discovered his "skull 1470," he submitted volcanic rock samples, from the stratum of the find, to F. J. Fitch and J. A. Miller, authorities on potassium-argon dating. That method gave a date of over 200 million years. Since no australopithecines could have lived so long ago, the dates were thrown out, the specimens considered tainted by "extraneous" argon. Fitch and Miller requested new samples, and dated them at approximately 2.6 million years. Subsequently *Nature* published a paleomagnetic study of fossils from the area, which affirmed: "An age of 2.7 to 3.0 Myr for this group is strongly indicated."[22] All this undoubtedly pleased Leakey, as it made him discoverer of the world's oldest human fossil.

Unfortunately, this caused a stir in paleoanthropology. Leakey's skull *looked* too modern for 2.6 million years. Pig fossils in the stratum did not correlate with that date either. Further radiometric tests were conducted; the earlier studies were criticized and invalidated; a new consensus put the skull at about 1.8 million years, a date "acceptable" to evolution.[23]

The point: While the public is led to believe radiometric techniques independently confirm evolution, these dates are rather plastic. If they agree with evolution, they are accepted; if not, they're tossed out. It's simple enough to call a date "bad" because of contamination, leaking or leaching. No proof is required that these distortions occurred; a "wrong" date means they did! A. Hayatsu, of the Department of Geophysics at the University of Western Ontario, noted:

> In conventional interpretation of K-Ar [potassium-argon] age data, it is common to discard ages which are substantially too high or too low compared with the rest of the group or with other available data such as the geological time scale. The discrepancies between the rejected and the accepted are arbitrarily attributed to excess or loss of argon.[24]

J. F. Evernden and John R. Richards stated in the *Journal of the Geological Society of Australia*:

> Thus, if one believes that the derived ages in particular instances are in gross disagreement with established facts of field geology, he must conjure up geological processes that could cause anomalous or altered argon contents of the minerals.[25]

Richard L. Mauger, associate professor of geology at East Carolina University, said:

> In general, dates in the "correct ball park" are assumed to be correct and are published, but those in disagreement with other data are seldom published nor are discrepancies fully explained.[26]

In *Radiocarbon Variations and Absolute Chronology*, T. Säve-Söderbergh and Ingrid U. Olsson, of the University of Uppsala, Sweden, reported on the Twelfth Nobel Symposium:

> C14 dating was being discussed at a symposium on the prehistory of the Nile Valley. A famous American colleague, Professor Brew, briefly summarized a common attitude among archaeologists toward it, as follows: "If a C14 date supports our theories, we put it in the main text. If it does not entirely contradict them, we put it in a footnote. And if it's completely 'out of date,' we just drop it."[27]

The polonium problem

Another radiometric phenomenon, ignored by evolutionists, suggests a young Earth. Robert V. Gentry is "the world's leading authority on the observation and measurement of anomalous radioactive haloes." That's how he was described by Francis Johnson, assistant director of the National Science Foundation's Division of Astronomical, Atmospheric, Earth, and Ocean Sciences.[28]

Gentry published numerous papers in *Science* and *Nature* on radioactive halos found in the Earth's deep granites. As a radioactive element decays, it emits energy in the form of particles or rays. This damages the mineral in which it is embedded, creating a halo. These are often quite beautiful. Each element produces a characteristic halo; by examining the halo, scientists can identify which element caused it.

Gentry became interested in the halos of polonium-218. This isotope has a half-life of only three minutes—it's barely around before it's gone. Yet it has left behind trillions of halos in the deep Precambrian granites of the Earth—the foundation rocks of the continents. This presents a problem. As Gentry notes:

> According to evolutionary geology, the Precambrian granites containing these special halos had crystallized gradually as hot magma slowly cooled over long ages. On the other hand, the radioactivity which produced these special radiohalos had such a fleeting existence that it would have disappeared long before the hot magma had time to cool sufficiently to form a solid rock.[29]

Critics protested that since polonium-218 is one of many isotopes formed during uranium's long decay into lead, it must have derived from uranium. But Gentry could find no traces of uranium—which also leaves a characteristic halo—in the polonium's vicinity. Professor R. G. Kazmann summed up in *Geotimes*:

> The polonium halos, especially those produced by ^{218}Po, are the center of a mystery. The half-life of the isotope is only 3 minutes. Yet the halos have been found in granitic rocks, at considerable depths below land sur-

> face, and in all parts of the world, including Scandinavia, India, Canada, and the United States. The difficulty arises from the observation that there is no identifiable precursor to the polonium; it appears to be primordial polonium. If so, how did the surrounding rocks crystallize rapidly enough so that there were crystals available ready to be imprinted with radiohalos by alpha particles from ^{218}Po?[30]

Gentry compares the situation to dropping an Alka-Seltzer tablet in water—its bubbles disappear in minutes. But if they were instantly frozen, the bubbles would be retained. Gentry believes the Earth and universe were created swiftly, just as Genesis says, and calls polonium halos "God's fingerprints." He has written a book summing his findings, *Creation's Tiny Mystery*.

Radiometric dating is a complex subject. This chapter has not intended to imply that radiometric scientists are dishonest, or their procedures disproven. However, I have attempted to illustrate that these methods are far more questionable than their public image conveys. As the preceding chapter recounted, many approaches can date the Earth besides radioactive decay, which is not uniquely authoritative. Radiometric techniques work—*assuming* decay rates have been stable, *assuming* no daughter elements existed in the original specimens, *assuming* no leaching or contamination has occurred, and, in the case of carbon dating, *assuming* atmospheric carbon ratios have never changed. Belief in these methods, and in an old Earth, requires a degree of *faith*.

Biologist Gary Parker, whom I like to quote, recounts a college experience that helped turn him from evolution to creation:

> In one graduate class, the professor told us we didn't have to memorize the dates of the geologic systems since they were far too uncertain and conflicting. Then in geophysics we went over all of the assumptions that go into radiometric dating. Afterwards, the professor said, "If a fundamentalist ever got hold of this stuff, he would make havoc out of the radiometric dating system. So, keep the faith." That's what he told us, "keep the faith."[31]

PLATE 40. Willard F. Libby, innovator of carbon dating

PLATE 41. Robert V. Gentry, leading authority on radioactive halos

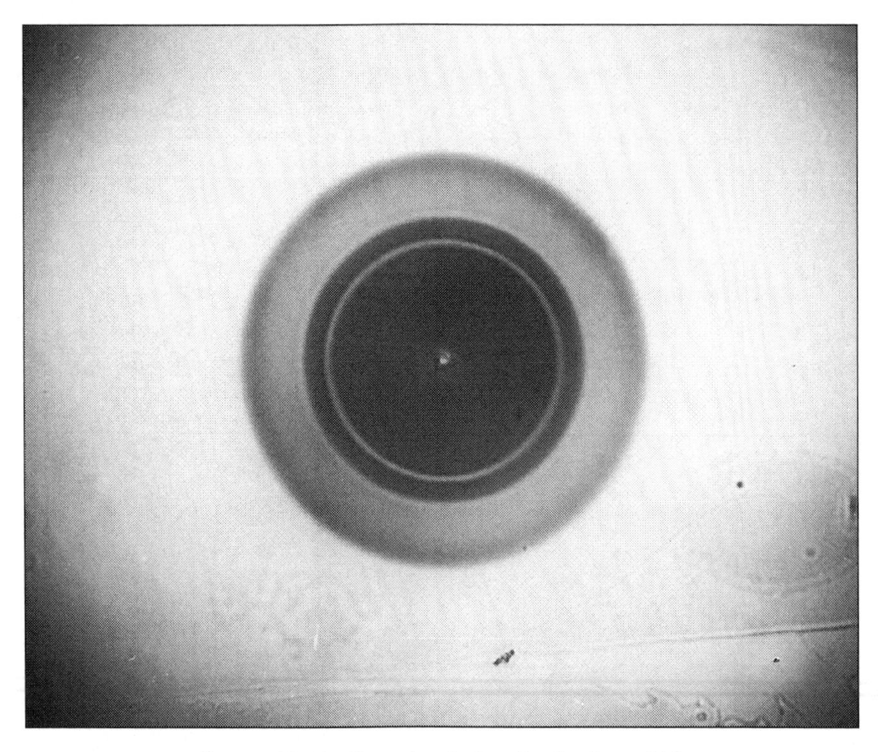

PLATE 42. Radioactive halo of polonium-218

CHAPTER 14

Rocks of Ages

Today, the Earth's layers are mapped in a familiar chart called the "geologic column." A version is on page 161. Each layer is assigned to a time period lasting millions of years. Strata are usually designated based on the fossils they hold. Many rocks containing dinosaurs, for instance, are placed in the Jurassic Period.

It has been said that creationists do not accept the evidence of geology. Actually, creationists do not differ with physical geology, only with the modern *interpretation* of what it means about the past.

The founders of the science—men like Nicolaus Steno and John Woodward—were creationists who considered the Flood of Noah the most significant geologic event in world history. Yale professor Benjamin Silliman (1779–1864), first president of the Association of American Geologists, said:

> With the Bible in my hands, and the world before me, I think I perceive a perfect harmony between science and revealed religion. . . . It cannot be doubted that there is a perfect harmony between the works and the word of God.[1]

The geologic column, in fact, was originally developed by creationists such as Adam Sedgwick, William Whewell, and William Coneybeare. Sedgwick, for example, who named the Cambrian and Devonian strata, was a devout Christian and ordained minister. Whewell, who named the Eocene, Miocene and Pliocene layers, was an Anglican priest. These men divided the geologic column into "systems," not "periods" or "eras"—the latter terms were appended later.

Most rock in the column, of course, is sedimentary—mainly deposited by water action. Until the 1800s, this rock, and the fossils therein, were considered remnants of the Flood (although the different layers suggested to some a *series* of catastrophes). The geologist who most transformed the field along evolutionary lines was England's Charles Lyell (1797–1875).

Working on a concept developed earlier by James Hutton, Lyell proposed that the geologic layers simply resulted from natural processes visible today—gradual erosion and sedimentation. In other words, sedimentary rock had been formed just as rain erodes a mountain, and a river leaves deposits in a delta—drip by drip, grain by grain. Lyell summed his views in his book *Principles of Geology*, subtitled *Being an Attempt to Explain the Former Changes of the Earth's Surface by Reference to Causes Now in Operation.*

Lyell's geology became known as "uniformitarianism": present processes have always been occurring, more or less uniformly, and account for nearly everything in the past. For example, shallow-water lime muds are now being deposited in tropical oceans at a rate of one foot per thousand years;[2] so it is assumed that limestone in places like the Grand Canyon also formed quite slowly, requiring eons.

Many natural laws have been formulated by observing current processes. But geology, unlike chemistry and physics, deals with past events that the scientist cannot reproduce. Geologic activity is not a constant, like the speed of light. It can occur slowly, but also *rapidly*, as during a flood, earthquake, or volcano. Lyell, however, minimized the impact of catastrophic events.

Fortuitously for evolution, the first volume of Lyell's book was published in 1830. When Charles Darwin departed on his famous cruise aboard the *Beagle* in 1831, he took a copy. Darwin later said: "I feel as if my books came half out of Sir Charles Lyell's brain."[3] Evolution could never work, of course, within the Biblical time frame. But here Lyell was saying the fossil-bearing rock layers had formed over millions of years. This made Darwin's hypothesis feasible.

After returning to England, Darwin became close friends with Lyell. He wrote: "I saw more of Lyell than of any other man, both before and after my marriage."[4] The geologist encouraged Darwin to publish *The Origin of Species* and later actively promoted it. As Lyell's uniformitarianism and Darwin's evolution seemed mutually corroborative, both theories grew in stature until they became the standards for geology and biology respectively. Neither man openly challenged the Bible, thus avoiding the stigma of admitting atheism, but their work's net effect was to invalidate the Scriptures. Darwin commented in 1873:

> Lyell is most firmly convinced that he has shaken the faith in the Deluge far more efficiently by never having said a word against the Bible than if he had acted otherwise . . . I have read lately Morley's *Life of Voltaire* and he insists strongly that direct attacks on Christianity (even when written with the wonderful force and vigor of Voltaire) produce little perma-

> nent effect; real good seems only to follow the slow and silent side attacks.[5]

Although there is a modern trend in geology toward accepting catastrophes as a partial explanation for geologic history, Lyell's uniformitarianism dominated the discipline for a century and a half.

How do geologists determine a rock stratum's age, and thus what sector of the geologic column to assign it to? As we have noted, radiometric techniques do not normally work on sedimentary rocks. (The geologic column, with its great time periods, had already been worked out before such methods were developed.) In most cases, geologists date a stratum simply by the fossils in it. They call these "index fossils." But how do they know how old the *fossils* are? Primarily, from Darwinian assumptions about when they evolved. Thus, the dating of the geologic column rests on some shaky ground. I quote a number of authorities:

Ronald R. West, assistant professor of paleobiology at Kansas State University:

> Contrary to what most scientists write, the fossil record does not support the Darwinian theory of evolution because it is this theory (there are several) which we use to interpret the fossil record. By doing so we are guilty of circular reasoning if we then say the fossil record supports this theory.[6]

Paleontologist Niles Eldredge of the American Museum of Natural History:

> And this poses something of a problem: If we date the rocks by their fossils, how can we then turn around and talk about patterns of evolutionary change through time in the fossil record?[7]

Tom Kemp, curator of zoological collections at Oxford University:

> A circular argument arises: Interpret the fossil record in the terms of a particular theory of evolution, inspect the interpretation, and note that it confirms the theory. Well, it would, wouldn't it?[8]

R. H. Rastall, lecturer in economic geology, Cambridge University:

> It cannot be denied that from a strictly philosophical standpoint geologists are here arguing in a circle. The succession of organisms has been determined by a study of their remains embedded in the rocks, and the relative ages of the rocks are determined by the remains of the organisms that they contain.[9]

J. E. O'Rourke in the *American Journal of Science*:

> The intelligent layman has long suspected circular reasoning in the use of rocks to date fossils and fossils to date rocks. The geologist has never

> bothered to think of a good reply, feeling that explanations are not worth the trouble as long as the work brings results.[10]

The public generally believes that the Earth's layers look much like the textbook geologic column, with strata neatly stacked upon each other. This is not the case. As geologist Steven A. Austin notes:

> The notion that the earth's crust has an "onion skin" structure with successive layers containing all strata systems distributed on a global scale is not according to the facts. Data from continents and ocean basins show that the ten systems are poorly represented on a global scale: approximately 77% of the earth's surface area on land and under the sea has *seven or more* (70% or more) of the strata systems *missing* beneath; 94% of the earth's surface has *three or more* systems *missing* beneath; and an estimated 99.6% has *at least one missing* system. Only a few locations on earth (about 0.4% of its area) have been described with the succession of the ten systems beneath (west Nepal, west Bolivia, and central Poland). . . . The entire geologic column, composed of complete strata systems, exists only in the diagrams drawn by geologists![11]

The Grand Canyon, which offers the best view of strata, contains only five of the major systems.[12] Evolutionists try to explain missing layers by saying they must have eroded away. But signs of erosion are often lacking. And if uniformitarianism is true, and erosion is almost always slow, how could it eliminate rock layers formed over tens of millions of years?

Another problem: in hundreds of locations, geologic strata exist in a different order than listed in the geologic column. *Science News* has noted:

> In many places, the oceanic sediments of which mountains are composed are inverted, with the older sediments lying on top of the younger.[13]

Examples include the famed Matterhorn, or the Mythen Peaks of the Alps, where Triassic Rock (supposedly some 200 million years old) rests atop Eocene Rock (40–50 million years old); or the Heart Mountain Thrust of Wyoming, which, as Whitcomb and Morris noted:

> occupies roughly a triangular area, 30 miles wide by 60 miles long, with its apex at the northeast corner of Yellowstone Park. It consists of 50 separate blocks of Paleozoic strata (Ordovician, Devonian, and Mississippian) resting essentially horizontally and conformably on Eocene beds, some 250,000,000 years younger![14]

Evolutionists attempt to explain discrepancies by saying one layer was shoved over another by crust activities—"overthrusts," "folding," faults, landslides. But while some cases of inverted strata display geologic evidence for this, others do not.[15]

Again, geologists usually identify rocks as "Devonian," "Triassic," etc., by their fossils. But fossils frequently do not appear in the order evolution expects. Marine fossils have been found in mountains around the world—even near the top of Mount Everest. In many places, complex life forms are found beneath the simple. As Austin notes:

> Although it is possible to find a series of fossils in the [Grand] Canyon and elsewhere to be in the same order of evolution, the pervasive pattern is contrary. Sponges, foraminifera, and corals, which are thought to precede trilobites in evolution, are first found in the Canyon *above* the first trilobites. Brachiopods, which are thought by many to be evolutionary descendants of bryozoa, are found *below* the first bryozoa, both in the Canyon and worldwide. Finally, recent studies have placed the trilobites as the most evolutionarily advanced of the arthropods, yet they are the first arthropods to be found, both in the Canyon and elsewhere in the world.[16]

Another problem for the geologic column is *polystrate fossils*—single fossils occupying more than one layer ("poly" means "multiple" and "strate" strata). For example, coal supposedly took eons to form. Yet upright trees are commonly found in coal seams, extending through "millions" of geologic years. How so? The trees would have rotted long before deposits could have accumulated around them in the slow, uniformitarian scenario.

New scientific studies are showing that coal, oil, and natural gas do not require an eternity to form, as previously thought. *Science News* has reported that Australian researchers have artificially produced a "wet natural gas,"[17] and that Exxon scientists have accelerated the generation of oil from shales. The journal noted of the latter:

> [T]he Exxon group collected samples of oil shale from different parts of the world. . . . then placed the samples into a pressurized reaction vessel and heated them individually to temperatures ranging from 570°C to 750°C. These hotter-than-natural conditions sped up the transformation from a geologic time frame of millions of years to one measured in days and hours.[18]

Research and Development reported:

> A group at Argonne National Laboratory near Chicago, IL, recently uncovered some clues as to the origin of coal. The studies indicate that currently accepted theories of the development of coal probably are wrong. . . . The Argonne team carried out its experiments at 300° F (150° C), a temperature that is fairly common in geological formations. The group heated undecomposed lignin, the substance that holds plants together, in the presence of montmorillite, or illite clay. The process led to

> simple coals, whose rank depended on the length of exposure to the 300° F temperature.[19]

Another difficulty for uniformitarian geology is bending strata. In sections of the Grand Canyon, for example, layers rest on each other in an undulating pattern (see picture and figure on page 166). If the rocks really formed over eons, they would have completely hardened before subsequent strata were imposed. How, then, could they later contort without tremendous fracturing? Such features *would* make sense, however, if the various layers had still been relatively soft when bent.

A better model

The Bible's worldwide Flood would explain much: why there are wavy strata; why most rock is sedimentary; why marine fossils are found in mountains.

It is true, of course, that while the fossil record lacks transitional forms, and often runs in the wrong order, there *is a trend*. One generally does find invertebrates at the lowest geologic levels, fish above them, then amphibians, reptiles, mammals, and finally man. This seems to support evolution.

But say there was a great Flood *today*: the continents overrun with water, land savagely eroded, massive sediments poured into the sea. What creatures would be buried lowest? Those that dwell at the lowest levels: clams and other invertebrates on the ocean floor.

In our imaginary scenario, who would get buried next? Well, who lives over the marine invertebrates? Fish. Their ability to swim above the sediments would exceed that of invertebrates.

As the oceans and lakes overflowed, and water rose on the continents, the next creatures buried would probably be those living at water's edge: amphibians. Then would come terrestrial animals. The ones with the least mobility for escape? Reptiles. Above them, one would find mammals; and at the greatest height, the most resourceful being of all: man.

Now, the fossils of these creatures would not always be in precise order, but this would probably be the trend. And that is just what we see in the geologic column. Do rocks reveal evolution's history—or the sequence of ecological zones affected by the Flood of Noah?

Another question. Countless billions of fossils are in the Earth. But today, millions of fish die daily, and, with very rare exceptions, they don't leave any fossils. Neither do other animals. Why not?

Most fossils are bones mineralized by sediments. A dead fish may float, like a goldfish in a bowl, or it may sink. It is either devoured by predators, or decomposes—but rarely is it buried under sediment before one of these events occurs. This is true of land animals as well. Millions of buffaloes were killed in the last century, but they left no fossils—scavengers and decomposing bacteria took care of that. Fossilization means *quick burial.*

One evidence for this is preservation of soft body parts. Millions of fish fossils include fins and scales. In fact, many fossils are of *entirely* soft-bodied creatures, such as jellyfish and worms. Even the fronds of palm trees have been preserved. Obviously, such organisms were buried fast. We even have fossils of animals giving birth and eating each other (see photos on p. 162). Anna K. Behrensmeyer noted in *American Scientist*:

> Once an organism dies, whether by attrition or catastrophe, there is usually intense competition among other organisms for the nutrients stored in its body. This combined with physical weathering and the dissolution of hard parts soon leads to destruction unless the remains are quickly buried. . . .
>
> These mechanisms contrast with the popular image of burial as a slow accumulation of sediment through long periods of time, a gentle fallout from air or water that gradually covers organic remains.[20]

That "popular image" is still seen at Harvard University's Museum of Cultural and Natural History, where a graphic depicts fossil formation: a horse dies on riverbank; the river floods while the carcass is fresh; sediment buries it; then, after millions of years, erosion exposes the fossil.

But the museum's uniformitarian "one-animal-at-a-time" explanation is belied by the billions of fossils found packed together in immense graveyards around the world. The *Journal of Paleontology* noted:

> Robert Broom, the South African paleontologist, estimated that there are eight hundred thousand million skeletons of vertebrate animals in the Karroo formation.[21]

Harold S. Ladd of the U.S. Geological Survey wrote in *Science* that in California's Monterey shale, "more than a billion fish, averaging 6 to 8 inches in length, died on 4 square miles of bay bottom."[22] England's Old Red Sandstone has areas with over a thousand fossil fish per cubic yard. That such formations denoted massive and sudden death was recognized even in the nineteenth century, when geologist Hugh Miller remarked:

> At this period of our history, some terrible catastrophe involved in sudden destruction the fish of an area at least a hundred miles from boundary to boundary, perhaps much more. The same platform in Orkney as at Cromarty is strewed thick with remains, which exhibit unequivocally the marks of violent death. The figures are contorted, contracted, curved; the tail in many instances is bent round to the head; the spines stick out; the fins are spread to the full, as in fish that die in convulsions.[23]

Dinosaurs, too, have graveyards. In Utah, for example, forty to sixty allosaurs were found buried together in 1940. The skeletons were disarticulated, suggesting a disaster befell them. The absence of bones of small juveniles implied they may have been fleeing. In fact, the fossilized bones of most animals are found higher than their fossilized footprints,[24] suggesting flight upward, as one would expect in a flood.

At the Harvard University fossil exhibit previously mentioned, there is a collection of rhinoceros bones excavated in Nebraska. The display sign reads:

> One of the interesting features of the deposit is its composition. As this section shows, the bones do not lie in any natural association, but are disarticulated and intermingled. . . . It is difficult to reconstruct the circumstances which led to the formation of this deposit. Perhaps a local catastrophe, possibly a severe drought. . . .

Even uniformitarians have to admit the role of catastrophes, but they invariably call it "local," since to do otherwise could make the Bible right. For if the geologic column is the result of Noah's Flood, the theory of evolution goes down the tubes! The truth is, however, that sudden and massive death is the mark of fossils worldwide, suggesting a disaster worldwide.

Upheaval in geology

Derek Ager, former President of the British Geologists' Association, said in a 1993 book:

> Uniformitarianism triumphed because it provided a general theory that was at once logical and seemingly "scientific." Catastrophism became a joke and no geologist would dare postulate anything that might be termed a "catastrophe" for fear of being laughed at or (in recent years) linked with a lunatic fringe of Velikovsky and Californian fundamentalists. But I would like to suggest that, in the first half of the last century, the "catastrophists" were better geologists than the "uniformitarians."[25]

A breakthrough eventually came in 1923, when geologist J. Harlen Bretz proposed that an ancient flood had sculptured numerous channels found in eastern Washington State. Later he extended the claim to include the steep, fifty-mile trench of the Grand Coulee. He cited facts demonstrating that the natural dam of a prehistoric body of water, Lake Missoula, had broken loose, creating these and other erosional features.

Bretz's conclusions were rejected for forty years, because they didn't conform to Lyell's uniformitarian dogma of gradual erosion. But Bretz continued pressing his case, and evidence accumulated. Finally, in 1965, after examining the land features, members of the International Association for Quaternary Research sent Bretz a telegram saying, "We are now all catastrophists."[26]

With time and observation, it has become clear that geologic transformations result more from natural disasters than drop-by-drop, grain-by-grain processes. The island of Surtsey, off the coast of Iceland, was initially formed in 1963 by a volcano in just a few hours.[27]

Concrete doesn't require millennia to harden; neither do sediments. In 1980, the eruption of Mount St. Helens created strata up to 600 feet high; within five years they had hardened into rock.[28] The volcano toppled 150 square miles of forest in six minutes. Its mud flows eroded a canyon system as deep as 140 feet; later on, a small creek was found running through it.[29]

Had this eruption occurred a couple of centuries ago, a modern uniformitarian geologist, inspecting these features, might conclude that the strata had formed over eons, and that the creek had carved the canyon.

This is the approach still taken to what most regard as nature's greatest wonder, the Grand Canyon. Supposedly the Colorado River cut the Canyon over millions of years. But that is not proven. As R. J. Rice commented in *Geographical Magazine*: "After a century of study, we seem, if anything, to be further than ever from a full comprehension of how the Grand Canyon has evolved."[30] Earle E. Spamer of Philadelphia's Academy of Natural Sciences noted in 1989:

> The greatest of Grand Canyon's enigmas is the problem of how it was made. . . . Grand Canyon has held tight to her secrets of origin and age. Every approach to this problem has been cloaked in hypothesis, drawing upon the incomplete empirical evidence of stratigraphy, sedimentology, and radiometric dating.[31]

According to legends of the Havasupai Indians, who still dwell in the Canyon, it was formed after a great flood covered the world.[32] Evidence is mounting that the Canyon, much like Bretz's Washington channels, was shaped when the natural dams of two ancient lakes gave way. An enormous, saucer-shaped basin exists east of the Canyon. If the Colorado River was blocked at the Canyon, a 30,000-square-mile lake would fill this basin.[33] Steven A. Austin's book *Grand Canyon: Monument to Catastrophe* makes a good case for flood formation of this famous landmark.

Lake Missoula, as well as the hypothetical lakes that created Grand Canyon, may have been bodies of water left over from the Great Flood, that eventually breached their boundaries.

Clearly, much of Earth's geologic history is a matter of interpretation. A jigsaw puzzle is relatively easy if you have a picture to go by—difficult without one. Today, more geologists are accepting catastrophism, ending the uniformitarian monopoly. In the *Journal of Geological Education*, Edgar B. Heylmun declared in his article "Should We Teach Uniformitarianism?":

> The fact is, the doctrine of uniformitarianism is no more "proved" than some of the early ideas of world-wide cataclysms have been disproved.[34]

Stephen Jay Gould wrote in *Natural History*:

> Charles Lyell was a lawyer by profession, and his book is one of the most brilliant briefs ever published by an advocate. . . . Lyell relied upon two bits of cunning to establish his uniformitarian views as the only true geology. First, he set up a straw man to demolish. . . . In fact, the catastrophists were much more empirically minded than Lyell. The geologic record does seem to record catastrophes: rocks are fractured and contorted; whole faunas are wiped out. . . . To circumvent this literal appearance, Lyell imposed his imagination upon the evidence. . . .
>
> Secondly, Lyell's "uniformity" is a hodgepodge of claims. One is a methodological statement that must be accepted by any scientist, catastrophist and uniformitarian alike. Other claims are substantive notions that have since been tested and abandoned. Lyell gave them a common name and pulled a consummate fast one. . . .[35]

Warren D. Allmon stated in *Science* in 1993:

> As is now increasingly acknowledged, however, Lyell also sold geology some snake oil. He convinced geologists that because physical laws are constant in time and space and current processes should be consulted first before resorting to unseen processes, it necessarily follows that all past processes acted at essentially their current rates. . . . This extreme gradualism has led to numerous unfortunate consequences, including the rejection of sudden or catastrophic events in the face of positive evidence for them, for no reason other than that they were not gradual.
>
> Indeed, geology appears at last to have outgrown Lyell. In an intellectual shift that may well rival that which accompanied the widespread acceptance of plate tectonics, the last 30 years have witnessed an increasing acceptance of rapid, rare, episodic, and "catastrophic" events.[36]

Derek Ager, who is certainly no creationist, affirms: "The hurricane, the flood or tsunami [tidal wave] may do more in an hour or a day than the ordinary processes of nature have achieved in a thousand years."[37] And if the hurricane, local flood or tsunami could accomplish so much, what then, the Flood of Noah?

TIME SCALE	ERAS	DURATION OF PERIODS	PERIODS	EPOCHS	DOMINANT ANIMAL LIFE
	CENOZOIC 70 MILLION YEARS DURATION		Quaternary	Recent Pleistocene	Man
10 20 30 40 50 60 70		70	Tertiary	Pliocene Miocene Oligocene Eocene Paleocene	Mammals
80 90 100	MESOZOIC 120 MILLION YEARS DURATION	50	Cretaceous		Dinosaurs
		35	Jurassic		
150		35	Triassic		
200	PALEOZOIC 350 MILLION YEARS DURATION	25	Permian		Primitive Reptiles
		20	Pennsylvanian		
250		30	Mississippian		
300		65	Devonian		Amphibians
350		35	Silurian		Fishes
400		75	Ordovician		Invertebrates
450 500		90	Cambrian		
Figures in millions of years	PROTEROZOIC ARCHAEOZOIC	Figures in millions of years	1500 MILLION YEARS DURATION		BEGINNINGS OF LIFE

PLATE 43. An older version of the geologic column (most periods are now considered to extend somewhat further back in time).

FOSSILIZATION MEANS

PLATE 44. This fish was buried before it finished breakfast.

PLATE 45. *Ichthyosaur* fossilized while giving birth

SWIFT BURIAL

PLATE 46. Dinosaur mass-burial bed at Dinosaur National Monument, Utah. Fossil "graveyards" indicate simultaneous, rapid death.

PLATE 47. Fish entombed together

PLATES 48 & 49. Soft matter, such as insect wings, leaves, and crinoids (plate 50, below left), were fossilized before decay set in.

PLATE 50

PLATE 51. Charles Lyell, founder of uniformitarian geology

PLATE 52. These strata, rapidly formed by catastrophic processes from Mount St. Helens, convey an appearance of great age.

PLATE 53. Mudflows from Mount St. Helens helped to cut this new canyon. Note the stream running through it.

PLATE 54. Folded sandstone in the Grand Canyon. If the strata were fully hardened when flexed, there should have been much more fracturing than is evident. For scale, note the people near bottom of photograph.

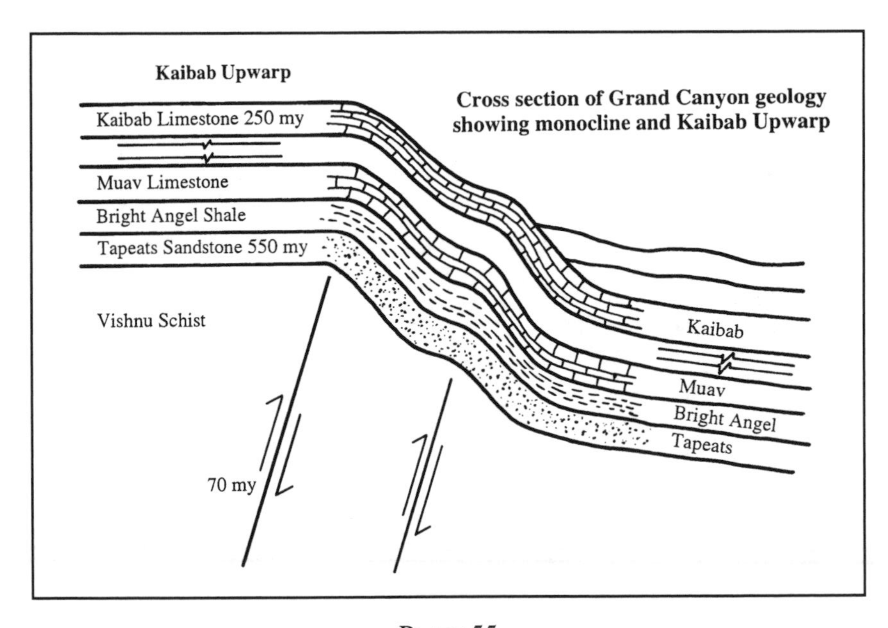

PLATE 55

CHAPTER 15

The Flood Remembered

One of the most compelling evidences for the Flood of Noah lies not in rocks but the world's cultures. More than two hundred have had legends pertaining to a great Flood.[1] For example, stories of a deluge were long widespread among Native Americans. After Congress commissioned a comprehensive study of Indian tribes in 1847, H. R. Schoolcraft reported:

> There is one particular in which the tribes identify themselves with the general traditions of mankind. It is in relation to a general deluge, by which races of men were destroyed. The event itself is variously related by an Algonquin, an Iroquois, a Cherokee, a Muscogee, or a Chickasaw; but all coincide in the statement that there was a general cataclysm, and that a few persons were saved.[2]

The ancient Greeks knew of the great Flood; both Aristotle and Plato referred to it. Flood legends also existed in the ancient cultures of Ireland, Wales, Norway, Lithuania, Romania, Egypt, Sudan, Syria, Persia, India, Russia, China, Indonesia, Polynesia, Hawaii, and Mexico; it was told of by ancient Celts and Incas; aboriginal tribesmen of Formosa and Indians of the Aleutian Islands; and scores of other cultures, great and small.

The Bible says God brought a flood upon the Earth because he saw "how great man's wickedness on the earth had become and that the inclination of the thoughts of his heart was only evil all the time." To reconcile the story with modern geology, which condones only regional catastrophism, some have interpreted the Biblical Flood as a local Middle East event. But the Bible clearly describes the Deluge as global, saying that the waters "rose greatly on the earth, and all the high mountains under the entire heavens were covered," and that "Every living thing that moved on the earth perished."[3] God saved Noah, however, who was righteous, as well as his wife, his three sons and their wives—eight people in all—by having them build and stay in the ark. And of course, God instructed Noah to bring "two of all living creatures, male and female, to keep them alive with you."

According to the Bible, water covered Earth for about a year. When the Flood receded, the ark came to rest on the mountains of Ararat. Noah kept releasing birds to see if the waters had withdrawn enough. One day, a dove came back with an olive leaf in its beak; a week later, it no longer returned. "Then Noah knew that the water had receded from the earth." Noah, his family, and the animals then left the ark, and began repopulating the Earth. And God created rainbows as a reminder of his new covenant with Noah.

In 95 percent of the more than two hundred flood legends, the flood was worldwide; in 88 percent, a certain family was favored; in 70 percent, survival was by means of a boat; in 67 percent, animals were also saved; in 66 percent, the flood was due to the wickedness of man; in 66 percent, the survivors had been forewarned; in 57 percent, they ended up on a mountain; in 35 percent, birds were sent out from the boat; and in 9 percent, exactly eight people were spared.[4]

Duane Gish sums up the Hawaiian tradition:

> Long after the death of Kuniuhonna, the first man, the world became a wicked, terrible place to live. There was one good man left; his name was Nu-u.
>
> He made a great canoe with a house on it and filled it with animals. The waters came up over all the earth and killed all the people. Only Nu-u and his family were saved.[5]

And he describes a Chinese legend:

> Ancient Chinese writings refer to a violent catastrophe that happened to the Earth. They report that the entire land was flooded. The water went up to the highest mountains and completely covered all the foothills. It left the country in desolate condition for years after.
>
> One ancient Chinese classic called the "Hihking" tells the story of Fuhi, whom the Chinese consider to be the father of their civilization. This history records that Fuhi, his wife, three sons, and three daughters escaped a great flood. He and his family were the only people left alive on earth. After the great flood they repopulated the world.
>
> An ancient temple in China has a wall painting that shows Fuhi's boat in the raging waters. Dolphins are swimming around the boat and a dove with an olive branch in its beak is flying toward it.[6]

According to Chinese tradition, Fuhi lived from 2852 to 2738 B.C.—about the time of Noah's Flood. His life span, 114 years, accords with the longevity the Bible assigns to ancient times.

A better-known Chinese story is that of Fuhi's sister, Nu-wah, which seems to combine several images from Genesis. Nu-wah, who had the face of a woman and body of a snake, formed the first people out of dirt—striking reminders of the Garden of Eden. According to Chinese legend, the sky broke, causing a flood, and Nu-wah stopped it by sewing up the sky with gemstones which formed a rainbow.

Edgar A. Truax translated the legend of inner China's Miao tribe, a separate people from ethnic Chinese. Their account resembles Genesis vividly. After God created the world, it says, the Earth became evil, and he resolved to destroy humanity. Picking up the Miao account:

> So it poured forty days in sheets and in torrents,
> Then fifty-five days of misting and drizzle.
> The waters surmounted the mountains and ranges.
> The deluge ascending leapt valley and hollow.
> And earth with no earth upon which to take refuge!
> A world with no foothold where one might subsist!
> The people were baffled, impotent and ruined,
> Despairing, horror stricken, diminished and finished.
> But the Patriarch Nuah was righteous.
> The Matriarch Gaw Bo-lu-en upright.
> Built a boat very wide.
> Made a ship very vast.
> Their household entire got aboard and were floated,
> The family complete rode the deluge in safety.
> The animals with him were female and male.
> The birds went along and were mated in pairs.
> When the time was fulfilled, God commanded the waters.
> The day had arrived, the flood waters receded.
> Then Nuah liberated a dove from their refuge,
> Sent a bird to go forth and bring again tidings.
> The flood had gone down into lake and to ocean;
> The mud was confined to the pools and the hollows.
> There was land once again where a man might reside;
> There was a place in the earth now to rear habitations.[7]

Since Mr. Truax was a missionary, many atheists may disbelieve his translation. But the universality of Flood accounts is undeniable. Some try to discount the legends as entirely the result of missionaries spreading the Bible. But these stories preceded Christianity's arrival, and even if they had come from missionary indoctrination, why, then, did they *differ* from the Biblical account in various ways, and why didn't the cultures know other Bible stories as well?

They didn't know those other stories, of course, because the Bible, after recounting the Flood, primarily describes God's relationship with the Jewish people, and the life of Christ, which took place in Israel. Other ancient peoples, who spread to distant parts of the Earth after the Flood, would not have known of these matters, but would certainly remember the Flood itself. Through centuries of mostly oral tradition, the Flood story was altered and embellished, accounting for the various discrepancies. In general, the further a culture is from Ararat, where the ark came to rest, the more the stories vary from the Biblical version.[8]

Anthropologists today say civilization began in the Middle East—just where one would expect if the ark settled there. The oldest historical documents are found in the Middle East.

I know that a host of objections exist regarding Noah's ark. Let's consider the major ones.

Come on, gimme a break! There's millions of species out there. You seriously expect anyone to believe they all fit in the ark?

As we noted in Chapter Five, "species" is a loose term. Varieties, such as Darwin's finches, are often mistaken for species. In 1967, about a hundred finches, all alike, were transported from the Pacific island of Laysan to a distant, isolated cluster of four islands. When examined twenty years later, the birds were found to look different from the original group, and also varied between islands, especially in bill shape.[9] This did not require the eons evolutionists attribute to the development of Darwin's finches. And obviously it was not due to mutations, which are infinitesimally rare events. Rather, the change came from the finches' *inherent genetic capacity* to vary and adapt. (This doesn't mean, of course, that birds could turn into cats or other animals, which would exceed their genetic limits.)

The sweet pea has been bred from a single type into hundreds of varieties since the year 1700.[10] By the same token, it would not have been necessary for Noah to have had a dalmatian, beagle, terrier, bulldog, and collie. Even evolutionists believe the hundreds of dog breeds originated from very few common ancestors. The primordial dogs undoubtedly had substantial capacity for genetic variation. But as some dogs became geographically isolated, or artificially bred, genetic information was lost, and they became distinguished types. These same phenomena would be true of other animals.

So Noah didn't need "millions of species" on the ark—only the progenitors of the innumerable varieties seen today. Furthermore, the ark was gigantic—according to the Bible, 300 cubits long (about 450 feet), 50 cubits wide, and 30 cubits deep. Its estimated cargo capacity was equal to 522 modern box cars,[11] more than enough to carry all the animals. Only a few creatures, such as elephants, are quite large, and even these could have been juveniles.

The Bible says it rained for forty days. Well, I got news for ya, Jack. The atmosphere can't hold that much water. And suppose the Earth really was covered with water above the mountains. Where'd it all go? Did God pull a big plug?

As we previously noted, Genesis One indicates that when Earth was created, there were waters above, which some creation scientists interpret as a canopy of water vapor. This may have been released at the time of the Flood, when the Bible says "the floodgates of the heavens were opened." The canopy would not have equated to today's atmosphere and rain capacity.

But the Bible first identifies another source of Flood waters, perhaps far more significant. It says there was "water under" when Earth was created, and that when the Deluge came, "all the springs of the great deep burst forth."[12] Of course, even today, springs yield water, and deep-sea explorations have revealed they are common to the ocean floor. But the Bible's description is one of great quantity and force, likely implying volcanic activity. Few volcanoes are active now, but tens of thousands of extinct ones exist, touching every area of the globe, including thousands on the Pacific floor. The Flood combined waters from above and below.

Where did the water go? The Earth's original surface was probably relatively smooth. The Bible says that, after the Flood, "the mountains rose; the valleys sank down."[13] That may be when most mountain-building—a process still poorly understood—occurred, and the ocean basins formed. It helps explain why mountains today have so many marine fossils. Where did the water go? Take a walk down to the ocean sometime. You'll be looking at it. Seventy-two percent of Earth's surface is covered with water.

OK, Einstein, if you're so smart, how did all the animals get to the ark? What, did kangaroos hop from Australia? Did polar bears walk there from the North Pole?

Even if we disregard supernatural intervention, we must remember that the world's topography was not the same in Noah's day. The continents probably did not exist as we now see them. Furthermore, as we have mentioned, palm tree fossils have been found in Alaska and fossils of tropical leaves in Antarctica. Much other evidence shows the ancient world had lush vegetation and a rather uniform climate. Camel fossils have been found in Oregon,[14] hippopotamus fossils in England and Germany,[15] and rhinoceros fossils in Siberia.[16] A 90-foot fruit tree, its fruit and leaves intact, has been excavated from the frozen ground of the New Siberian islands.[17]

Even the Gobi Desert was once rich with life. As Michael J. Novacek et al. reported in *Scientific American*:

> The Gobi Desert of Central Asia is one of the earth's most desolate places. . . . Yet the Gobi is a paradise for paleontologists. . . . dinosaurs, lizards and small mammals in an unprecedented state of preservation. . . . Among them are 25 skeletons of therapod dinosaurs . . . more than 200 skulls of mammals . . . an even greater number of lizard skulls and skeletons.[18]

Conditions before the Flood were apparently ideal. If there was a water canopy, it would have exercised a warming greenhouse effect, and perhaps added oxygen to the environment. Fossils we find today—giant beavers, giant dragonflies, etc.—indicate a better state for living. Cosmic radiation would have been reduced by the then-stronger magnetic field (see Chapter Twelve) and the water canopy's shield; thus, mutations would have been much rarer. These factors may explain the great human longevity the Bible notes for ancient times: centuries before the Flood, with rapidly decreasing life spans after the Flood, until finally, in King David's time, we read: "The length of our days is seventy years—or eighty, if we have the strength." (Psalm 90:10)

Noah's kangaroos did not have to hop from Australia—although most marsupials live only there today, their fossils are found on several continents.[19] Nor would polar bears have come from the North Pole—an icy Arctic would not have existed. Most of Noah's animals probably lived nearby. Polar bears may just be a variety who migrated north after the Flood, and adapted to newer, colder conditions.

Panda bears eat only bamboo, koala bears only eucalyptus leaves. Did Noah really have supplies of that?

If he didn't, these dietary "necessities" may have resulted from adaptation to isolated environments after the Flood. And a number of animals with unique dietary habits have been known to adjust when their normal food became unavailable.[20]

Answer this, Bozo. How could all the world's races come from Noah and his family?

The Bible says eight people were on the Ark: Noah, his wife, three sons, and three daughters-in-law. The daughters-in-law could have come from diverse racial backgrounds. When people think of race, they typically think of skin color, which is just a matter of how much melanin or pigment you have—an inherited trait based on genes. When a black person marries a white person, they ordinarily give birth to mulattos, who are medium-brown in skin color. When two mulattos marry, their children may be totally black, totally white, or brown-skinned in color—it just depends on the gene combination inherited. In other words, two mulattos could give birth to a race of black people and a race of white people, and it would require only one generation, not eons. The Bible says that after the Flood, the peoples separated and migrated. People with more pigment can better tolerate sunlight; this is probably why darker-skinned people tended to settle in warmer climates, and lighter-skinned people in colder ones.

If one argues that eight people could not genetically produce the world's races, then Darwin is in trouble. According to evolution, one primordial cell produced everything from blue whales to palm trees to ladybugs to people.

John Woodmorappe has written a scholarly book, *Noah's Ark: A Feasibility Study*, dealing with food storage, waste disposal, and many other issues relating to the ark. It is available from the Institute for Creation Research, PO Box 2667, El Cajon, CA 92021, (619) 448-0900.

Myth versus memory

As discussed in Chapter Twelve, we would get the current population from the ark's eight survivors, if the annual growth rate since had been only 0.5 percent (today's growth rate is around 1.6 percent). The Bible, incidentally, lays out a table of nations in Genesis 10. It lists the three sons of Noah (Japheth, Ham and Shem) and their descendants, who became the tribes, races and nations of the world. The eminent archaeologist William F. Albright of Johns Hopkins University said:

> The tenth chapter of Genesis has long attracted students of ancient Oriental geography and ethnography. It stands absolutely alone in ancient literature, without a remote parallel, even among the Greeks, where we find the closest approach to a distribution of peoples in genealogical framework. . . . The Table of Nations remains an astonishingly accurate document.[21]

Based on 25 years of research, British author Bill Cooper has written a book called *After the Flood*. Utilizing ancient documents, it validates the accuracy of the Genesis table of nations. The ancient Jews, of course, who were descendants of Shem, kept meticulous genealogies still found in the Bible's Old Testament. Less well known, however, is that many other tribes kept similar records. As Cooper notes:

> [T]hroughout the entire Table of Nations, whether we talk about the descendants of Shem, Ham or Japheth, every one of their names is found in the records of the early surrounding nations of the Middle East, even many of the obscure names of certain remote Arab tribes that are otherwise not evident in any modern history book of the times, and enough is available for a detailed history to be written about them.[22]

Among Cooper's sources on the Middle East were Arab historians; the works of Ptolemy, Josephus, and other ancient writers; and Babylonian, Assyrian and Egyptian records.

Just a small sampling of the genealogies Cooper documents: Asshur, whom the Bible lists as a son of Shem, founded Assyria; Madai, a son of Japheth, began the people known as the ancient Medes; the modern Lebanese port of Saida is built on the site of ancient Sidon, named for Zidon, a descendant of Ham who established the Phoenician people; Jectan, a town in Saudi Arabia, perpetuates the name of Joktan, descendant of Shem and father of several Arab tribes; Dodanim, whom Genesis names as a son of Japheth, originated a people that the Greeks called the Dardanians—his name still remembered in the Dardanelles strait. Moslems, of course, trace their roots to Ishmael, a descendant of Shem. Genesis 25 lists Hadad, still a common Arabic name, as Ishmael's son.

Our point: Noah's descendants, listed in the Bible's table of nations, are not "some old myth." Ancient histories and modern names preserve them. It sure beats trying to explain everything by distribution of ape ancestors.

Cooper also takes a fascinating excursion into the history of the European peoples, proving that they considered themselves descendants of Japheth.

Little is commonly said of the Britons prior to Caesar's invasion in 55 B.C. Britain's conquest by various cultures—Romans, Saxons, Normans—suppressed the remembrance of early British history. But ancient records still existed, and were summarized in the eighth century by the Welsh historian Nennius in his book *History of the Britons*, and in the twelfth century by the Welsh historian Geoffrey of Monmouth, in *History of the Kings of Britain.* Nennius and Geoffrey traced the early British kings back to Brutus, a descendant of Japheth, the son of Noah. Brutus landed on Britain around 1100 B.C.

Most modern historians ignore or even ridicule the works of Nennius and Geoffrey as fictions. Cooper has found surviving medieval manuscripts of early Welsh chronicles at the National Library of Wales, Aberystwyth; the Free Public Library, Cardiff, Wales; Jesus College Library, Oxford; and the British Museum, London. He comments:

> The above list of chronicles that give the history of the early Britons, constitute a rather large percentage of the total number of Welsh manuscripts that have come down to us from medieval times. Given that they are all catalogued in easily accessible collections, it is astonishing that even their very existence goes unmentioned by most scholars who are aware of them, and that British history prior to 55 B.C. remains a blank page. But perhaps their acknowledgment would lead the recorded history of the early Britons uncomfortably back to Genesis, and that is a concept that modernism simply could not accommodate.[23]

The six royal houses of the Anglo-Saxons also traced their descent to Japheth;[24] but again, historians discount it as myth. Kenneth Sisam, writing in *Anglo-Saxon Royal Genealogies*, said:

> The Biblical names show the artificial character of this lengthened pedigree and the crudeness of the connexions that passed muster. Otherwise they need not detain us.[25]

Incredibly, evolution has so "discredited" Genesis that many take mere mention of Biblical names in an historical document as evidence of its forgery. But, as Cooper points out, ancient peoples revered their ancestors—tampering with genealogies would have been a serious matter to them.

The ancient Irish, though a distinct people from the ancient British, also traced their lineage to Japheth. Eventually they branched off. Irish genealogies were recorded by Ernín, son of King Duach, who died in A.D. 365, before St. Patrick brought Christianity to the island. The original chronicles, now lost, were preserved through such later books as Keating's *The History of Ireland* (1630).

The Norwegian and Danish kings also followed their ancestry back to Japheth. Odin, a famous figure in Norse mythology, was actually an ancestor who became deified—a seventeenth-generation descendant of Japheth.[26]

Most ancient societies worshiped multiple deities. However, a remembrance of one God, an Eden-like paradise, and other elements from Genesis could often be found in their cultures. According to the Greeks, the first woman, Pandora, opened a jar which brought evil upon mankind. Comparison to Eve is hard to avoid. Xenophanes, Greek philosopher of the sixth century B.C., stated:

> There is one God, greatest among gods and men, similar to mortals neither in shape nor in thought . . . . He sees as a whole, he thinks as a whole, he hears as a whole. . . . Always he remains in the same state, changing not at all. . . . But far from toil he governs everything with his mind.[27]

The Chinese connection

Earlier we recounted the Chinese legends of Fuhi and Nu-wa, which so resemble portions of the Old Testament. In their book *The Discovery of Genesis*, C. H. Kang and Ethel Nelson made a case that Chinese culture, especially its unique pictographic language, contains images rooted in the Bible's first chapters.

China's culture and writing are believed to have originated about 2500 B.C. Its three major religions—Buddhism, Confucianism, and Taoism—did not exist prior to 500 B.C. So who did the Chinese worship during the 2,000 years *before*?

Originally, the Chinese revered one god—*Shang Ti*, which means "emperor of heaven" and sounds similar to the old Hebrew word for God, *Shaddai*. Like the ancient Jews, the Chinese emperors offered sacrifices to heaven—in fact, the ritual continued until imperial rule ended in the twentieth century. John Ross noted in his 1909 book *The Original Religion of China*:

> One of the most honorable prerogatives of the Chinese emperor is the annual sacrifice offered at the winter solstice to the Supreme Ruler of the universe. Even four thousand years ago this sacrifice was a long-established practice, and the duty of performing it belonged of right to the reigning Sovereign. . . . The central act of this sacrifice consists of burning on an altar in the open air the entire body of a young bull as a burnt-offering to God.[28]

Part of the emperor's recitation during the ritual resembled the language of Genesis:

> Of old in the beginning, there was the great chaos, without form and dark. The five elements [planets] had not yet begun to revolve, nor the sun and the moon to shine. In the midst thereof there existed neither form nor sound. Thou, O spiritual Sovereign, camest forth in Thy presidency, and first didst divide the grosser parts from the purer. Thou madest heaven; Thou madest earth; Thou madest man. All things with their reproducing power, got their being. . . . Thou didst produce, O Spirit, the sun and the moon and the five planets. . . .[29]

Like Jews and Christians, the ancient Chinese regarded God as "a Father." In the ritual, the emperor continued:

> Thou hast vouchsafed, O Te, to hear us, for Thou regardest us as a Father. I, thy child, dull and unenlightened, am unable to show forth my dutiful feelings. . . . Honourable is Thy Great Name.[30]

After thousands of years, however, the ritual's meaning had been completely lost—little more than an "old custom." Ross pointed out that Confucius said that "the man who could explain the sacrifice to God would be able to rule the Empire as easily as he could look on the palm of his own hand."[31] (Traditions with forgotten purposes are not unusual. In America, the significance of Christmas and Easter—the birth and Resurrection of Christ—have largely become overshadowed by Santa Claus and the Easter Bunny.)

Genesis prescribed the traditional seven-day week, which, unlike the day and year, has no basis in any astronomical phenomenon. Despite isolation from other cultures for thousands of years, the ancient Chinese also had a seven-day week for a while. This again raises the question of common religious roots.

But the most intriguing link between China and the Bible may be language itself. Unlike English, which forms words from an alphabet, ancient Chinese consisted of thousands of pictographs—each written word being a likeness of the thing represented. Over time, the language has been somewhat modified, especially by adding phonetic elements to the pictures. In *The Discovery of Genesis*, Kang and Nelson used ancient pictographs, since these more truly reveal the language's original meaning.

The Chinese character for "boat" is 船, which combines the symbols for "vessel" 舟, "eight" 八, and "mouth" 口. Chinese often uses "mouth" as a synonym for "person," as in the English idiom, "many mouths to feed." Thus the Chinese symbol for boat is a vessel with eight people, just as the ark carried eight persons—Noah, his wife, three sons and their wives.

Kang and Nelson enumerated dozens of other Chinese words suggesting a relationship to Genesis. While the interpretation was to an extent subjective, they made a thought-provoking case. In 1997, Dr. Nelson, Dr. Ginger Tong Chock and Richard E. Broadberry released *God's Promise to the Chinese*, a book which updated the study using oracle bone characters, the most ancient Chinese writing known.

Was there a great Flood? Evolutionists will allow for *other* catastrophes—like a big asteroid wiping out the dinosaurs. But the Flood? No way! Ironically, during the 1997 Pathfinder mission to Mars, *Science* reported:

> But team scientists were already doing impressionistic science on images from the site, finding evidence that it was swept by one of the largest floods—or mud slides—in solar system history. . . .
>
> Golombek and his colleagues believe that the first images confirm their suspicion that billions of years ago a great flood of a billion cubic meters per second swept the region for weeks, carrying a variety of rocks from distant highlands.[32]

So get this, folks. Mars is dry—it has no known water (save tiny traces of water vapor in its atmosphere, and it has polar caps of undetermined composition). Nevertheless, scientists say it once had a "great flood." More than 70 percent of Earth, on the other hand, is covered with water. But a great flood here? Nah, impossible!

After all, that would agree with the Bible.

Plate 56. The story of Noah's ark is echoed in worldwide flood legends.

Plate 57. Floods inflict devastating damage. When California's St. Francis Dam broke in 1928, a wall of water swept the San Francisquito Canyon, killing nearly five hundred people. Nothing was left of the dam but the remnant above. Controversy surrounding the disaster later helped inspire the movie *Chinatown*.

CHAPTER 16

Dinosaurs, Dragons and Ice

On the old *Tonight* show, if Johnny Carson had just finished reading from the previous chapter, Ed McMahon would say to him: "Well, I guess that's it . . . That really said it all . . . I mean, there's nothing more that you could say about old legends. Everything—I mean EVERYTHING you could possibly know about old cultures, their myths and legends is right there in that chapter!" And Carson would say, "You are wrong, brontosaurus breath!"

The Flood is not the only common remembrance of the world's cultures. They also remember "dragons." From England to China, these were long a part of national "mythologies." The Indians of North and South America had legends about them. They were written of in Ireland, France, Germany, Italy, Greece, Switzerland, Scandinavia, Ethiopia, Egypt, Persia, Russia, India, and Japan.

Ever notice the uncanny resemblance between dragons and dinosaurs?—large reptiles with big teeth, long tails, plated backs? And like dinosaurs, dragons came in many descriptions and were reported on every continent. But hey, hey, I guess that's just another coincidence. I mean, it's a big coincidence that all those old cultures recalled the Flood—it never happened—and it's also a coincidence, them talking about dragons. Obviously, they imagined the beasts. And the similarity to dinosaurs? Well, ah, that's just people's biological subconscious. Peter Dickinson notes: "Carl Sagan tried to account for the spread and consistency of dragon legends by saying that they are fossil memories of the time of the dinosaurs, come down to us through a general mammalian memory inherited from the early mammals, our ancestors, who had to compete with the great predatory lizards."[1]

Memories of 65 million years ago, to be precise. That's supposedly when dinosaurs went extinct. Today, we consider just 2,000 years ago to be "ancient times." So when a paleontologist declares a dinosaur bone 100 million years old, the numbers are hard to relate to. Am I really saying dinosaurs were contemporary with man? If Earth is young, as this book proposes, they must have been.

"Oh, fiddlesticks! There may be some old legends about knights fighting monsters, but they didn't call them 'dinosaurs.'"

That's right. Because the word was unknown until the nineteenth century. Coined by British anatomist Richard Owen after studying some dinosaur bones, it means "terrible lizard." Other modern terms, like *Tyrannosaurus rex*, didn't exist before then either. In the meantime, ancient peoples had to call the creatures something, and they used "dragon."

According to the Roman historian Livy, a great dragon in Africa attacked the army of General Regulus. Livy wrote:

> After many of the soldiers had been seized in [the dragon's] mouth, and many more crushed by the folds of its tail, its hide being too thick for javelins and darts, the dragon was at last attacked by military engines and crushed by repeated blows from heavy stones.[2]

Time-Life Books' *The Enchanted World of Dragons* notes:

> The creatures' monstrous strength and ferocity was such that in classical times they were said to be predators of that other natural behemoth, the elephant: In Ethiopia, it was reported, dragons were called, simply, elephant killers.[3]

Were these great African reptiles imaginary? The *Los Angeles Herald Examiner* reported in 1970:

> A fantastic mystery has developed over a set of cave paintings found in the Gorozomzi Hills, 25 miles from Salisbury [in Rhodesia, now Zimbabwe]. For the paintings include a brontosaurus—the 67-foot, 30-ton creature scientists believed became extinct millions of years before man appeared on earth.
>
> Yet the bushmen who did the paintings ruled Rhodesia from only 1500 B.C. until a couple of hundred years ago. And the experts agree that the bushmen always painted from life.
>
> This belief is borne out by other Gorozomzi Hills cave paintings—accurate representations of the elephant, hippo, buck and giraffe.[4]

Far more evidence originated in Europe. Ken Ham et al. write:

> [I]n the tenth century, an Irishman wrote of his encounter with what appears to have been a *Stegosaurus*. In the 1500s, a European scientific book, *Historia Animalium*, listed several animals which to us are dinosaurs, as still alive. A well-known naturalist of the time, Ulysses Aldrovandus, recorded an encounter between a peasant named Baptista and a dragon whose description fits that of the dinosaur *Tanystropheus*. The encounter was on May 13, 1572 near Bologna, Italy, and the peasant killed the dragon.[5]

Aldrovandus also reported, in *Historia Serpentium et Draconium:* "Cracus [c. A.D. 700], who later named the city of Cracow [Poland] after himself, slew an immense dragon which laired in the aforementioned crag."[6]

Time-Life Books' *The Enchanted World of Dragons* notes that "Marco Polo reported seeing some lindworms [two-legged dragons] while crossing the steppes of Central Asia."[7] The book cites a number of other examples:

> **Ireland**: According to legends, Tristan of Lyonesse slew a dragon here in the 11th century. . . .
>
> **Provence, France**: A dragon called the Drac inhabited the Rhone River throughout the 13th century; the town of Draguignan was named for it, although its worst attacks seem to have occurred in Beaucaire. . . .
>
> **Isle Ste. Marguerite, France**: This island off the French coast sheltered a dragon during much of the Middle Ages; because of the beast's ferocity, it was often confused with the Tarasque. . . .
>
> **Sanctogoarin and Neidenburg, Germany**: The naturalist Edward Topsell wrote in 1608 that Sanctogoarin was plagued by a dragon whose flights caused fires; the dragon of Neidenburg poisoned wells by bathing in them. . . .
>
> **Bonn, Germany**: The Italian naturalist Ulisse Aldrovandi had in his collection a lindworm killed near Bonn in 1572. . . .
>
> **Switzerland**: Christopher Schorer, the Prefect of the canton of Solothurn, reported the sighting of a winged mountain dragon near Lucerne in 1619, as well as an encounter in 1654 between a hunter and a dragon. The latter retreated with a rustling of scales into its mountain den. . . .
>
> **Rome, Italy**: The Historia naturalis of Pliny the Elder reported that a dragon killed on Vatican Hill during the reign of the Emperor Claudius (died 54 A.D.) contained the body of a child; centuries later, in 1660, the German Athanatius Kircher examined a dragon killed near the city. He commented on its unusual webbed feet. . . .
>
> **Kiev, Russia**: As recorded in the byliny—legends of heroes—dating from the 11th Century, a dragon called Gorynych terrorized this region for years before the hero Dobrýnja slew it. . . .[8]

Ancient Persian kings were famed as dragon slayers. A Chinese emperor of the Sung dynasty was reputed to have raised one in his palace compound.[9]

Describing an area in China that has produced many dinosaur bones, D. Leland Niermann wrote in the *Creation Ex Nihilo Technical Journal*:

> The oldest record of possible dinosaur bones is in a Chinese book written between A.D. 265 and 317. It mentions "dragon bones" found at Wucheng, in Sichuan Province. . . . To the Chinese, "Dragon bones" and "Dinosaur bones" were one and the same.[10]

Author Rhoda Blumberg noted:

> Until 1927 entire villages in China supported themselves by digging and selling dragon bones. Visiting scientists from the American Museum of Natural History were very upset when they saw warehouses filled with the bones, because they claimed that these were fossils of prehistoric animals. They bought and shipped huge quantities to their museum in New York City, and presented some to a museum in Peking. Thereafter, selling dragon-bone medicine was forbidden. However, an undercover dragon-bone business continues to cater to customers all over the Orient.[11]

Perhaps England, with its tales of St. George and Sir Lancelot, is most famous for dragon legends. Geoffrey of Monmouth recorded how King Morvid died in c. 336 B.C.:

> For a beast, more fell than any monster ever heard of before, came up from the Irish sea and preyed continually upon the seafaring folk that dwelt in those parts. And when Morvid heard tidings thereof he came unto the beast and fought with her single-handed. But when he had used up all his weapons against her in vain, the monster ran upon him with open jaws and swallowed him up as [if] he had been a little fish.[12]

Granted, some legends have glaringly fictitious elements: multi-headed dragons, for example, or ones that guarded treasures. But true stories often develop embellishments. Even the tales of dragons that breathed fire could have an analogue in the bombardier beetle (Chapter Four), which sprays its enemies with an explosive mixture. Fossils of one class of dinosaur, the *Hadrosaurs* or "duck-billed" dinosaurs, show hollow protuberances on their heads, connected to their noses by tubes. The function of these structures is unknown.[13]

Many reports appear matter-of-fact. Researcher Bill Cooper found dragon accounts at nearly 200 sites in his native England. He relates, for example:

> Later in the fifteenth century, according to a contemporary chronicle that still survives in Canterbury Cathedral's library, the following incident was reported. On the afternoon of Friday, 26th September, 1449, two giant reptiles were seen fighting on the banks of the River Stour (near the village of Little Conard) which marked the English country borders of Suffolk and Essex. One was black, and the other "reddish and spotted." After an hour-long struggle that took place "to the admiration of many beholding them," the black monster yielded and returned to its lair, the scene of the conflict being known ever since as Sharpfight meadow.[14]

Or take the following chronicle, dated 1405, from Suffolk, England:

> Close to the town of Bures, near Sudbury, there has lately appeared, to the great hurt of the countryside, a dragon, vast in body, with a crested head, teeth like a saw, and a tail extending to an enormous length. Having slaughtered the shepherd of a flock, it devoured many sheep. . . . in order to destroy him, all the country people around were summoned. But when the dragon saw that he was again to be assailed with arrows, he fled into a marsh or mere and there hid himself among the long reeds, and was no more seen.[15]

The evolutionist will call such reports fiction. Our ancestors were, after all, superstitious morons, more closely related to apes, and imagined things. But suppose the above account was not of a dragon, but a wolf or other common beast? In that case, it would be accepted with little question.

The epic poem *Beowulf*, considered the first great work of English literature, describes a battle between the hero, Beowulf, and a monster called Grendel. Of Anglo-Saxon origin, it is not about Britain, which it never mentions, but Scandinavia, the story itself predating the Saxon migration to the Isles. Containing many mythic aspects, it has been dismissed as a fairy tale. However, it accurately mentions numerous historical figures listed in Anglo-Saxon, Danish and Swedish genealogies. Beowulf himself lived from A.D. 495 to 583 and was king of a tribe known as the Geats.

The monster's name, "Grendel," probably derived from Norse *grindill*, meaning storm, or *grenja*, meaning bellow.[16] It may have described a species, rather than merely the specific beast of the poem, who preyed upon the Danes for twelve years, A.D. 503 to 515, stalking the moors. All attempts to slay him had failed, as he was impervious to weapons. *Beowulf* described Grendel as having powerful jaws, with which he killed and ate many warriors. He stood on two feet, but evidently had weak forelimbs (Beowulf slew him by tearing one off—the creature then returned to its lair and bled to death). Sound like anyone you met in *Jurassic Park*?

The forearms were evidently the monster's vulnerable point. Cooper says "it is the very size of Grendel's jaws which paradoxically would have aided Beowulf in his carefully thought out strategy of going for the forelimbs, because pushing himself hard into the animal's chest between those forelimbs would have placed Beowulf tightly underneath those jaws and would thus have sheltered him from Grendel's terrible teeth."[17] (We wouldn't recommend trying this at home, but remember it if you ever run into a *T. rex*.)

Speculative? Highly. But if Beowulf really slew a *T. rex*, it would explain his fame as the subject of an epic poem. According to that saga, incidentally, Beowulf himself ultimately died battling a flying dragon.

Flying dragons? Oh, come on, man, that kills it as myth right there!

Does it? There were flying reptiles—the best-known being pterodactyls, who had wing spans of up to 39 feet. The fossil bones of one flying reptile, *Quetzalcoatlus*, unearthed in Texas in 1972, had a wing span of 48 feet—wider than that of an F-4 Phantom jet.

American Indians long had legends of great flying beasts, which they represented in petroglyphs. Brad Steiger noted in *Worlds Before Our Own*:

> Whatever the petroglyphs truly represented, all the Amerindian nations of what then constituted the Northwest Territory had a terrible tradition associated with the creatures they called *The Piasa* (or Piusa).[18]

According to Indian lore, the *Piasa* would swoop down and take tribesmen. P. A. Armstrong wrote in his 1887 booklet *The Piasa or the Devil Among the Indians* that the Indian petroglyphs portrayed the *Piasa* as having "the wings of a bat, but of the shape of an eagle's."[19] He further noted:

> The time when the Piasas existed in this country, according to the Illini tradition, was "many thousand moons before the arrival of the palefaces," while the Miamis says, "several thousand winters before the palefaces came." Though indefinite as to the exact time period, both indicate a very long period of time—many centuries—and may be construed to go away back to the mesozoic or middle-life geological period, known as the age of reptiles, when the monster saurians existed in great numbers and varieties. . . .
>
> Among the most notable were the pterodactyl, or wing-fingered monstrosity, which in every point of the horrible surpassed the ichthyosaur and plesiosaur. It was an aerial beast, bird, or reptile, with wings shaped like those of a bat. . . . The fossil remains of some twenty-five species of this monster have been found, and it is sometimes called the pterosaur or flying lizard. . . .
>
> Our conclusions may be summed up in a few words, as follows:
>
> First. The Indians appeared upon this continent before the extinction of the huge reptiles and saurians of the mesozoic age.

> Second. That among the still existing saurians or reptiles when the Indians appeared was one huge monster that could walk, run, fly, and swim, known to the Indians as the Piasa, whose bones have been found and reconstructed into the saurian, or reptile, known to science as the ramphorhyneus.
>
> Third. That this saurian or reptile was of immense size, great strength and voracious appetite. . . .[20]

Reports of flying reptiles persisted until the nineteenth century. We cite just a few examples. Time-Life Books' *The Enchanted World of Dragons* states that on November 30, 1222, flying dragons were seen over London; their flight was associated with thunderstorms. The same book notes that in Henham, in Essex, England, "An amphiptère [winged serpent] 9 feet long was discovered on a hillock near the town in 1669. The terrifying serpent remained in the area for some months but inflicted no actual harm. . . ."[21]

In *The Statistical Account of Scotland*, published in 1793, we read:

> In the end of November and beginning of December last, many of the country people observed very uncommon phenomena in the air, (which they call dragons), of a red fiery color, appearing in the north and flying rapidly towards the east, from which they concluded, and their conjectures were right, that a course of loud winds, and boisterous weather would follow.[22]

Here is a piece that appeared in *The Illustrated London News* of February 9, 1856:

> A discovery of great scientific importance has just been made at Culmont (Haute Marne). Some men employed in cutting a tunnel which is to unite the St. Dizier and Nancy railways, had just thrown down an enormous block of stone by means of gunpowder, and were in the act of breaking it to pieces, when from a cavity in it they suddenly saw emerge a living being of monstrous form. This creature, which belongs to the class of animals hitherto considered to be extinct, has a very long neck, and a mouth filled with sharp teeth. It stands on four long legs, which are united together by two membranes, doubtless intended to support the animal in the air, and are armed with four claws terminated by long and crooked talons. Its general form resembles that of a bat, differing only in its size, which is that of a large goose. Its membranous wings, when spread out, measure from tip to tip 3 metres 22 centimetres (nearly 10 feet 7 inches). Its colour is a livid black; its skin is naked, thick, and oily; its intestines only contained a colourless liquid like clear water. On reaching the light this monster gave some signs of life, by shaking its wings, but soon after expired, uttering a hoarse cry. This strange creature, to which may be given the name of living fossil, has been brought to Gray, where a naturalist, well

> versed in the study of paleontology, immediately recognized it as belonging to the genus *Pterodactylus anas*, many fossil remains of which have been found among the strata which geologists have designated by the name of lias. The rock in which this monster was discovered belongs precisely to that formation the deposit of which is so old that geologists date it more than a million of years back.[23]

On April 26, 1890, Arizona's *Tombstone Epitaph* carried this report:

> A winged monster, resembling a huge alligator with an extremely elongated tail and an immense pair of wings, was found on the desert between the Whetstone and Huachuca mountains last Sunday by two ranchers who were returning home from the Huachucas. The creature was evidently greatly exhausted by a long flight and when discovered was able to fly but a short distance at a time. After the first shock of wild amazement had passed the two men, who were on horseback and armed with Winchester rifles, regained sufficient courage to pursue the monster and after an exciting chase of several miles succeeded in getting near enough to open fire with their rifles and wounding it. The creature then turned on the men, but owing to its exhausted condition they were able to keep out of its way and after a few well directed shots the monster partly rolled over and remained motionless. The men cautiously approached, their horses snorting with terror, and found that the creature was dead. They then proceeded to make an examination and found that it measured about ninety-two feet in length and the greatest diameter was about fifty inches. The monster had only two feet, these being situated a short distance in front of where the wings were joined to the body. The head, as near as they could judge, was about eight feet long, the jaws being thickly set with strong, sharp teeth. Its eyes were as large as a dinner plate and protruded about halfway from the head. They had some difficulty in measuring the wings as they were partly folded under the body, but finally got one straightened out sufficiently to make a measurement of seventy-eight feet, making the total length from tip to tip about 160 feet. The wings were composed of a thick and nearly transparent membrane and were devoid of feathers or hair, as was the entire body. The skin of the body was comparatively smooth and easily penetrated by a bullet. The men cut off a small portion of the tip of one wing and took it home with them. Late last night one of them arrived in this city for supplies and to make the necessary preparations to skin the creature, when the hide will be sent east for examination by the eminent scientists of the day. The finder returned early this morning accompanied by several prominent men who will endeavor to bring the strange creature to this city before it is mutilated.[24]

The dimensions described in this latter story vastly exceed that of any fossilized flying reptiles, but comparison is hard to avoid. Was this a late survivor of the *Piasa* that terrorized Indian tribes?

Even the Bible has passages describing what many consider dinosaurs. In Job 40 we read:

> Look at the behemoth, which I made along with you, and which feeds on grass like an ox.
> What strength he has in his loins, what power in the muscles of his belly!
> His tail sways like a cedar; the sinews of his thighs are close-knit.
> His bones are tubes of bronze, his limbs like rods of iron.

Since Job describes many familiar animals, the behemoth was presumably real also. Some claim it was an elephant—but what elephant has a tail that swings like a cedar tree? A *Brachiosaurus* would. Later Job describes another unknown creature, the leviathan:

> Smoke pours forth from his nostrils as from a boiling pot over a fire of reeds.
> His breath sets coals ablaze, and flames dart from his mouth.
> Strength resides in his neck; dismay goes before him.
> The folds of his flesh are tightly joined; they are firm and immovable.
> His chest is hard as rock, hard as a lower millstone.
> When he rises up, the mighty are terrified; they retreat before his thrashing.
> The sword that reaches him has no effect, nor does the spear or the dart or the javelin.
> Iron he treats like straw, and bronze like rotten wood.[25]

Why did the dinosaurs become extinct? The most popular theory today: a big asteroid struck Earth, creating climatic changes that wiped them out. But why would an asteroid destroy all dinosaurs, large and small, yet not mammals and other reptiles?

Dinosaurs may have died out for the same reason as many other creatures—the filling of Earth by humans. Wild beasts were very prolific until man decimated them.

Tigers, for example, were once a bane in Asia. The thirteenth-century traveler Marco Polo wrote of Kue-Lin-Fu in China: "The multitude of tigers renders travelling through the country dangerous, unless a number of persons go in company." And he said of the Chintigui region: "In this province the tigers are so numerous, that the inhabitants, from apprehension of their ravages, cannot venture to sleep at night out of their towns. . . ."[26] But who today fears travel in China because of tigers? Now they are considered an endangered species.

Dinosaurs evidently went, unmysteriously, the way of other extinct beasts. Dragon slayers' exploits were recorded in the literature of country after country, and our ancestors would probably resent knowing we had relegated it all to fable and illusion.

A note on the ice age

But post-Flood climate changes may have also contributed to elimination of the dinosaurs. One major phase of Earth's past we have not yet discussed is the ice age.

Without question, ice sheets previously covered northern portions of Asia, Europe and America. As glaciers move, they scratch rocks, move debris, and leave other telltale signs. Ice sheets reached as far south in America as the Ohio and Missouri River valleys.

It is commonly claimed that there were many ice ages—classically, four—lasting around 100,000 years each. Today, twenty or even thirty ice ages are being proposed. This is primarily based on (1) some evidence of intermittent glaciation in the geologic layers; and (2) the assumption that geologic layers require eons to form. Freed from uniformitarian dogma, however, one ice age appears more likely, the intermittence resulting not from numerous ice ages, but the advance and retreat of a single ice sheet during climatic variations. Within the overall zone of past glaciation, there are "driftless" areas *showing no evidence of glaciers at all*, as in southwest Wisconsin.[27] This strongly weighs against multiple ice ages, as does the fossil record. Michael J. Oard notes:

> [N]early all of the extinctions of large mammals occurred after the last glaciation. Very few occurred after postulated, previous glaciations. Each ice age would have been very stressful to the animals. How could they survive 20–to–30 ice ages, over a two-or-three million year period, and then go extinct only after the last? The record of extinctions is more consistent with just one ice age.[28]

Dozens of theories attempt to explain the cause of the ice age(s). Since Earth supposedly began as a hot molten ball, and even now is believed warming, what could have occasioned massive glaciation in between? Colder winters alone wouldn't; we have cold winters now, and they don't generate ice sheets. Chilly summers would also be needed—otherwise the glaciers would melt.

Cold alone, however, wouldn't be enough; an ice age would require lots of snow. But ironically, colder air carries less moisture and yields less precipitation. Storms that occur just below freezing, of course, produce the heaviest snowfall. Somehow, the ice age must have entailed more moisture *and* colder temperatures—meteorological conditions nonexistent today.

The Bible provides a rational explanation. As the last chapter noted, the Flood would have accompanied eruption of the thousands of now dormant volcanoes, as the Biblical "waters below" gushed out and "all the springs of the great deep burst forth."

The Earth's interior is hot: temperatures increase by about 30 degrees centigrade for each kilometer down through the crust—so the "springs of the great deep" discharging would have produced warm oceans. Studies of oxygen isotopes in Antarctic sea microorganism deposits reveal that the ocean there was once warmer by some 20 degrees centigrade.[29] Warmer oceans mean more evaporation and precipitation; and computer projections at the National Center for Atmospheric Research have shown that they would cause especially heavy precipitation in polar regions.[30]

In the meantime, ash from all the volcanoes would have darkened the sky. Volcanic dust blocks sunlight, but does little to retain the Earth's heat, thus reducing atmospheric temperatures. The eruption of a single great volcano—Krakatoa in 1883—is believed responsible for a subsequent worldwide drop in temperatures that lasted five years. A similar effect occurred after Tambora's eruption in 1815: New England received six inches of snow in June 1816, and temperatures there went as low as 37 degrees Fahrenheit that August.[31]

The Flood would have thus created the exact combination—high precipitation with low temperatures—an ice age requires. Like today, equatorial regions would have been warmest, and polar areas coldest; but with average temperatures lower, glaciation would have extended further south than now. For although the oceans were warm, the land was not.

With so much water tied up in large terrestrial glaciers, the oceans would have been lower, creating land bridges between Siberia and Alaska (by which Indians reached the Americas) and probably down the East Indies as far as Australia. However, as volcanoes died down, and ash cleared from the atmosphere, temperatures would have climbed, making the glaciers recede and oceans rise.

During the ice age, the middle latitudes would have received more precipitation too, but as rain rather than snow. Areas now dry would have been nicely irrigated. I used to wonder why, in the Bible, God told the Jews he would bring them out of Egypt to "a land flowing with milk and honey," "a land with streams and pools of water, with springs flowing in the valleys and hills; a land with wheat and barley, vines and fig trees, pomegranates, olive oil and honey. . . ."[32] That didn't sound like the comparatively parched

Israel of recent years; I figured the Bible was way wrong. Fact is, when Moses led the Israelites out of Egypt, the region was more wet and fertile. Significant evidence shows that now-arid places, such as the deserts of Africa and the Western U.S., were once well-watered.[33] But precipitation there declined as the oceans cooled, evaporating less, and water locked up in northern glaciers.

Great ice layers do not require eons to form. During World War II, a squadron of eight warplanes left the U.S. Army air base in Greenland, headed for Britain. However, a massive blizzard forced them to turn back. Unable to touch down in Iceland as they had hoped, the squadron, running out of fuel, was forced to crash-land on an eastern Greenland ice sheet. All the crewmen were rescued. In 1988, an effort began to salvage the planes. It was assumed that, once the squadron was found, it would just be a matter of brushing the snow off. However, when radar located the planes, they were under approximately 250 feet of ice[34]—dispelling the uniformitarian idea that ice sheets develop over thousands of years. Incidentally, one of the planes, a P-38 Lightning fighter, was eventually dug out in rather good condition.

How long was the ice age? Certainly it was not fleeting, as many volcanoes would have continued to spew ash after the Flood. Meteorologist Michael J. Oard, who has written an insightful book, *An Ice Age Caused by the Genesis Flood,* estimates that it lasted about 700 years. People in northern climates probably wore animal skins for greater warmth then, and sought shelter in caves during the worst weather. The reduced sunlight during this period may be responsible for the Neanderthal skeletons with signs of rickets. All of these factors would have helped shape the modern misconception of so-called "cavemen."

The ice age may have also hastened dinosaur extinction. Most were plant eaters, and generally being large, ate great quantities. Thus their numbers would have dwindled during the ice age from inadequate food—that, of course, and the "dragon slayers."

PLATE 58. Medieval theory of dinosaur extinction

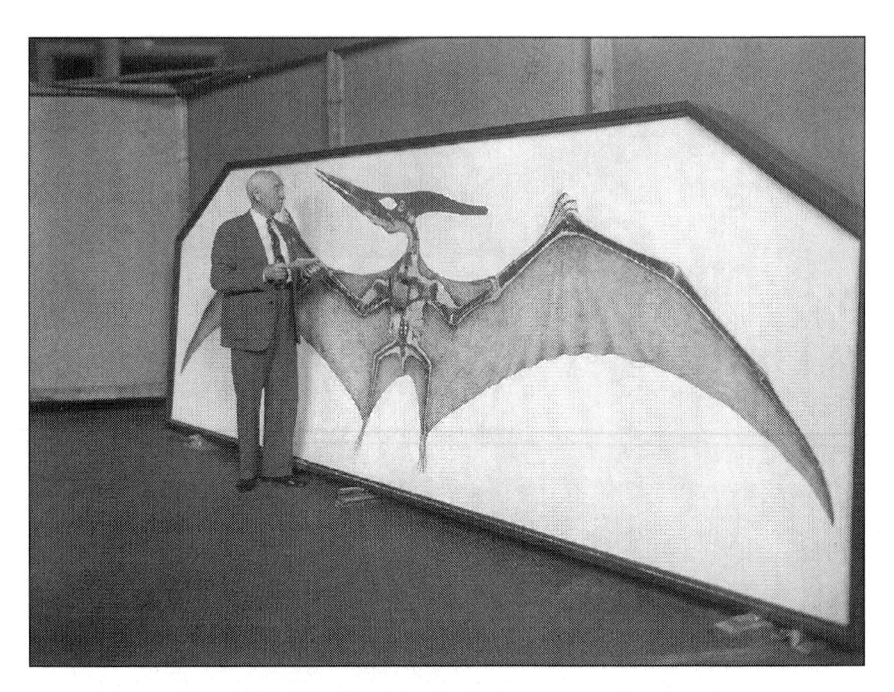

PLATE 59. The *Pteranodon*'s immense wing span

PLATE 60. A boy probes dinosaur tracks. Made when the ground was still wet, they supposedly remained undisturbed for tens of millions of years.

PLATE 61. Lowered temperatures from multiple volcanos could have contributed to an ice age.

CHAPTER 17

Trial by Hollywood

It has been said that fiction persuades people more effectively than non-fiction, because it does a better job of touching emotions. Perhaps nothing has advanced evolution's cause so effectively as a play and movie—*Inherit the Wind*.

Inherit depicts what was perhaps the most famous court case of the twentieth century—the "monkey trial" of 1925. The defendant was John Scopes, a schoolteacher from Dayton, Tennessee. He was charged with violating the Butler Act, a state law that forbade teaching that man descended from lower life forms (it did not prohibit teaching other aspects of evolution). The Butler Act had been uncontroversial in the Tennessee legislature, passing 71–5 in the house and 24–6 in the senate.

Leading Scopes's defense was Clarence Darrow, the most famous criminal lawyer of his day; assisting the prosecution was William Jennings Bryan, former Secretary of State and three times the Democratic Party's Presidential candidate. The most common impression about this trial is probably that Darrow humiliated Bryan in cross-examination, scoring a powerful blow for evolution against religious fundamentalism.

Public beliefs regarding the trial are based largely on *Inherit the Wind*. The play enjoyed a record three-year run on Broadway. It then became a film starring Spencer Tracy and Frederic March and was nominated for several Academy Awards. The movie, as well as a 1988 televised remake, have been shown countless times to students as "educational" material. The play has been frequently revived. Few people, however, have ever read the actual trial transcript. For most, *Inherit the Wind* IS the trial, and for many, even defines their perception of the creation-evolution debate.

It might be said, "Ah, come on, lighten up, nobody expects literal interpretations from Hollywood. Everyone knows that screenwriters sometimes change facts to make a story more interesting." That's right. I have signed away options on an unproduced screenplay of my own, and I know that when a writer fictionalizes a true event, he may have to create conflict where none existed, to put zip into the story, or invent new characters to generate dialogue.

That's not what I'm talking about. *Inherit the Wind* did not alter facts merely to stimulate the audience. It grossly perverted the Scopes trial to advance a specific agenda. It is true that the original playwrights, Jerome Lawrence and Robert E. Lee, acknowledged that their work "is not history,"[1] and changed the principals' names. John Scopes became Bert Cates; Clarence Darrow became Henry Drummond; William Jennings Bryan became Matthew Harrison Brady; Dayton, Tennessee became Hillsboro, Tennessee. Of course, everyone knew who they were talking about, but by making this disclaimer and changing the names, Lawrence and Lee padded their license to smear.

This chapter will contrast *Inherit the Wind* with the actual Scopes trial, by comparing the 1960 Spencer Tracy movie (the most familiar and accessible version) to the original courtroom transcript and other records.

IN THE MOVIE, the film opens as the grim town minister and other prudish-looking residents of Hillsboro gather. Ominous music plays against the hymn "Give me that old time religion, it's good enough for me." The citizens march to the local high school, where young Bert Cates (John Scopes) is forthrightly teaching biology, using Darwin's *Descent of Man.* Cates is portrayed as a man who grew up in Hillsboro; neighborhood children, we learn, would come to his house to peer through his microscope.

The town prudes arrest Cates on the spot. The arresting officer reads in a droning fashion from a warrant. Cates says to him: "Come off it, Sam, you've known me all my life."

IN REAL LIFE, John Scopes was not a biology teacher, nor did he grow up in Dayton, Tennessee. Scopes taught math and coached football, but had briefly substituted for the regular biology teacher during an illness. He was recruited by the American Civil Liberties Union (ACLU) to challenge Tennessee's Butler Act. Evidently, he never even taught evolution or Darwin. L. Sprague de Camp's *The Great Monkey Trial* relates the following conversation between Scopes and reporter William K. Hutchinson of the International News Service:

> "There is something I must tell you. It's worried me. I didn't violate the law."
>
> "A jury has said you had," replied Hutchinson.
>
> "Yes, but I never taught that evolution lesson. I skipped it. I was doing something else the day I should have taught it, and I missed the whole lesson about Darwin and never did teach it. Those kids they put on the stand

> couldn't remember what I taught them three months ago. They were coached by the lawyers. And that April twenty-fourth date was just a guess.
>
> "Honest, I've been scared all through the trial that the kids might remember I missed the lesson. I was afraid they'd get on the stand and say I hadn't taught it and then the whole trial would go blooey. If that happened, they'd run me out of town on a rail."
>
> "Well, you are safe now," said Hutchinson.[2]

Don't buy it? Here's what Scopes himself said in his autobiography, *Center of the Storm*:

> To tell the truth, I wasn't sure I had taught evolution.[3]
>
> Darrow had been afraid for me to go on the stand. Darrow realized that I was not a science teacher and he was afraid that if I were put on the stand I would be asked if I actually taught biology.[4]

And Scopes wrote of his students who were called as witnesses:

> If the boys had got their review of evolution from me, I was unaware of it. I didn't remember teaching it.[5]

IN THE MOVIE, we next see Cates in jail, where he learns that the famous Matthew Harrison Brady is coming to prosecute him. The jailer asks, "Who's gonna be *your* lawyer, son?" Cates replies: "I don't know yet. I wrote to that newspaper in Baltimore. They're sending somebody."

Cates's girlfriend, Rachel, pleads with him: "Tell them you're sorry. Tell them it was a mistake." But Bert says: "Tell them if they let my body out of jail, I'd lock up my mind? Could you stand that, Rachel?"

At night, having recently heard a fiery sermon by the town preacher, a mob gathers outside Cates's jail, burning him in effigy and threatening to lynch him. To the tune of "The Battle Hymn of the Republic," they sing: "We'll hang Bert Cates to a sour apple tree, our God is marching on! Glory, glory, hallelujah. . . ." A rock is hurled through the jail window, injuring him.

In the courtroom, Cates is told: "We'll fix you, Cates—we'll run you out of town!"

John Scopes is thus portrayed as a heroic martyr, persecuted by witch-hunting bigots for daring to speak the truth.

IN REAL LIFE, Scopes never spent one second in jail. Violating Tennessee's Butler Act was not an imprisonable offense; it was punishable only by a fine (which Scopes was never required to pay).

Furthermore, there was no bad blood between Scopes and Dayton's people. The entire affair was amicably arranged. The ACLU had been running ads in Tennessee newspapers, offering to pay expenses for any teacher who would volunteer to participate in a court challenge to the new law. (The ACLU, of course, is well known for opposing organized religion; it has frequently sued public schools that allow religious teachings and towns that display nativity scenes at Christmas.)

George Rappleyea, manager of a mining company, noticed the ad. He convinced local businessmen that such a trial would put Dayton on the map (which it did) and hopefully lift its sagging economy. The men approached Scopes in the local drugstore run by "Doc" Robinson. They asked John if he would agree to say he had violated the law and be served with a warrant. Scopes later recalled the conversation:

> "You filled in as a biology teacher, didn't you?" Robinson said.
>
> "Yes." I nodded. "When Mr. Ferguson was sick."
>
> "Well, you taught biology then. Didn't you cover evolution?"
>
> "We reviewed for final exams, as best I remember." To tell the truth, I wasn't sure I had taught evolution.
>
> Robinson and the others apparently weren't concerned about this technicality. I had expressed willingness to stand trial. That was enough.[6]

The Dayton businessmen were so eager to have the trial, that when they learned Chattanooga was trying to get its own court case going, Daytonians threatened to boycott Chattanooga merchants, and Scope's indictment was accelerated.[7]

To be sure, John Scopes believed in evolution. However, his trial was not instigated by witch-hunting fundamentalists, but by the ACLU, which not only paid the defense's costs, but offered to pay the prosecution's as well (an offer that was turned down). Everything happened with Scopes's consent. Far from lynching him, the townspeople gave Scopes a seat of honor next to William Jennings Bryan at a banquet held for the latter.

IN THE MOVIE, when the locals learn Henry Drummond (Clarence Darrow) is coming to be Bert Cates's defense attorney, they yell: "We'll send him back to hell!" "Ride him out on a rail!" "Don't let him into town! Keep him out!"

When he arrives, he is greeted by only one person—newspaperman E. K. Hornbeck (based on the cynical journalist H. L. Mencken). (Drummond was played by Spencer Tracy and Hornbeck by Gene Kelly.) Drummond gets a rough reception. A big, gruff farmer rebukes him. A senile-looking Bible salesman asks Hornbeck: "Are you an evolutionist? An infidel? A sinner?" The only hospitable folks are Bert Cates's enlightened students. When Drummond approaches the courthouse the next morning, he is loudly booed. At night, a mob of fundamentalists outside his hotel threaten to lynch him.

IN REAL LIFE, a friendly crowd greeted Clarence Darrow at the train station. The town held a banquet in his honor. Here is what Darrow himself said of his experience there:

> Yet I came here a perfect stranger and I can say what I have said before that I have not found upon anybody's part—any citizen here in this town or outside, the slightest discourtesy. I have been better treated, kindlier and more hospitably than I fancied would have been the case in the north, and that is due largely to the ideas that southern people have and they are, perhaps, more hospitable than we are up north.[8]

IN THE MOVIE, Henry Drummond is Bert Cates's lone attorney—an underdog fighting the system, represented by Matthew Harrison Brady, the state attorney, and a bigoted judge.

IN REAL LIFE, Clarence Darrow brought to Dayton a team of lawyers, including ACLU heavyweight Arthur Garfield Hays, New York divorce attorney Dudley Field Malone, and, for insight into local law, Tennessean John Neal.

IN THE MOVIE, in sharp contrast to Drummond's cold reception, Matthew Harrison Brady (William Jennings Bryan) is given a huge parade. Marching before him, singing "Gimme that old time religion, it's good enough for me," are the town's housewives, who, to a woman, are matronly, dour, prudish and frumpy (does that about sum it up?). They change the lyrics to: "If it's good enough for Brady, then it's good enough for me." On the spot, the mayor makes Brady an honorary colonel in the state militia.

In court, the judge prejudicially keeps referring to Brady as "Colonel." Drummond says: "And I object to all this 'Colonel Brady' talk. I am not familiar with Mr. Brady's military record." Smitten, the mayor reluctantly makes Drummond a colonel, too.

IN REAL LIFE, no parade was given for Bryan, who *was* a colonel in the U.S. Army during the Spanish-American War, though he saw no action. And the judge, John T. Raulston, courteously referred to Darrow as "Colonel" from their first courtroom exchange, with no wrangling over titles.

IN THE MOVIE, Brady is an ignorant bigot opposed to all science. He says, "The people of this state have made it very clear that they do not want this zoological hogwash slobbered around the schoolrooms!" He declares: "The way of scientism is the way of darkness."

IN REAL LIFE, Bryan was a member of the American Association for the Advancement of Science. Here is what he really said about science during the trial:

> The Christian men and women of Tennessee know how deeply mankind is indebted to science for benefits conferred by the discovery of the laws of nature and by the designing of machinery for the utilization of these laws. Give science a fact and it is not only invincible, but of incalculable service to man.[9]

IN THE MOVIE, Brady opposes evolution solely on Biblical grounds.

IN REAL LIFE, he *also* opposed it on rational, scientific grounds. In an article published in *Reader's Digest* Bryan said:

> It is not unusual for evolutionists to declare that their hypothesis is as clearly established as the law of gravitation or the roundness of the earth. Yet anyone can prove that anything heavier than air, when thrown up into the air, will fall to the ground; anyone can demonstrate the roundness of the earth by traveling around it.
>
> But how about the doctrine that all of the species . . . by the operation of interior, resident forces came by slow and gradual development from one or a few germs of life, which appeared on this planet millions of years ago—the estimates varying according to the vigor of the guesser's imagination and the number of ciphers left in his basket? . . . On the contrary, no one has ever been able to trace one single species to another. Darwin admitted that no species had ever been traced to another, but he thought his hypothesis should be accepted even though the "missing links" had not been found. . . . If there is such a thing as evolution, it is not just one link—the link between man and the lower forms of life—that is missing, but all the millions of links between millions of species. . . .
>
> When a few bones and a piece of skull are fashioned into a supposed likeness of a prehistoric animal, described as an ape-man, the evolutionists fall down before it and worship it, although it contains a smaller percentage of fact than the one-half percent alcohol permitted in a legal beverage. . . .[10]

IN THE MOVIE, Brady is completely unfamiliar with Darwin's works. After he declares his knowledge of the Bible, the following exchange ensues:

> Drummond: I don't suppose there are many portions of *this* book you've committed to memory—the *Origin of the Species*?
>
> Brady: I am not in the least interested in the pagan hypotheses of that book.
>
> Drummond: Never read it?
>
> Brady: And I never will.
>
> Drummond: Then how in perdition have you got the gall to whoop up this holy war about something that you don't know anything about?

IN REAL LIFE, Bryan quoted Darwin extensively, in both the Dayton courtroom and his writings.

IN THE MOVIE, the prosecution objects when the defense tries to introduce Darwin's texts as evidence; the bigoted judge agrees and excludes them.

IN REAL LIFE, not only were Darwin's books allowed in evidence, but *Bryan* introduced them. I quote the trial transcript:

> Mr. Bryan: Let me read what Darwin says. . . .
> Mr. Malone [defense attorney]: What is the book, Mr. Bryan?
> Mr. Bryan: "The Descent of Man," by Charles Darwin.
> Mr. Malone: That has not been offered as evidence?
> Mr. Bryan: I should be glad to offer it.[11]

IN THE MOVIE, Brady is an obnoxious boor who laughs at his own corny jokes and can never resist making long speeches in the courtroom. Even the judge at last seems exasperated by Brady's penchant for speeches.

IN REAL LIFE, Bryan never spoke a word in court until the fourth day of the trial, and that was in response to a query. The defense was quoting Bryan's writings on religious freedom; asked if he minded, he said: "Not a bit."[12]

In contrast, Clarence Darrow had already engaged in considerable oratory during the trial, including a two-hour speech on religion, bigotry and the law.

IN THE MOVIE, Drummond must turn down a bigoted juror, who proclaims: "I believe in Matthew Harrison Brady!" Brady accepts jurors based solely on their belief in the Bible, and even tries to renege when he learns that one is not as dogmatically religious as he had hoped.

IN REAL LIFE, Darrow excused some jurors, but not for such flagrant prejudice, and Bryan never spoke during jury selection.

IN THE MOVIE, in one of the script's worst misrepresentations, the judge disallows any testimony from eminent scientists whom Drummond has brought to the trial. In a droning voice, the judge declares that "zoology" (which he can barely pronounce) and other scientific topics are "irrelevant to the case."

IN REAL LIFE, Darrow called as a witness Maynard Metcalf, a zoologist from Johns Hopkins. He testified at length.[13] It is true that Judge Raulston excluded the *jury* from that testimony. Darrow had instructed Scopes to plead "not guilty"; the jury's only responsibility was to determine if he had broken the law. It was not their duty to decide if evolution was true, or if the Butler Act was constitutional.

It soon became clear that the atheistic Darrow was orchestrating a parade of witnesses for the purpose of promoting evolution. (The trial was being broadcast by radio across the nation, and reported in all the newspapers.) The prosecution correctly protested that this was irrelevant to the legal question at hand—Had Scopes violated the Butler Act?—and the judge, after studying the issue, concurred.

However, the defense argued that if Judge Raulston heard further scientific testimony, he would realize he was wrong. Giving great leeway, he courteously consented to hear more from the experts. Here is what he said:

> The Court—I am going to let you introduce evidence and I will sit here and hear it, and if that evidence were to convince me that I was in error I would, of course, reverse myself.

William Jennings Bryan then raised a point that riled Clarence Darrow:

> Mr. Bryan—I ask your honor: Will we be entitled to cross-examine their witnesses?
>
> The Court—You will, if they go on the stand.
>
> Mr. Darrow—They have no more right to cross-examine than to bring in the jury to hear this issue.

The judge then asked Darrow a pertinent question:

> The Court—Colonel, what is the purpose of cross-examination?
>
> Mr. Darrow—The purpose of cross-examination is to be used on the trial.
>
> The Court—Well, isn't it an effort to ascertain the truth?
>
> Mr. Darrow—No, it is an effort to show prejudice.[14]

Obviously, when one side in a trial calls witnesses, the opposing party has a right to cross-examine them. But Darrow knew Bryan would ask his experts tough questions like "Where are the missing links?" Even worse, he might ask if they were atheists—which some could not deny without perjuring themselves. All this would spoil Darrow's evolutionary showcase. He decided to have his witnesses instead make written affidavits for submission to an appeals court, thus avoiding any risk of cross-examination:

> Mr. Darrow—We expect to protect our rights in some other court. Now, that is plain enough, isn't it? Then we will make statements of what we expect to prove. Can we have the rest of the day to draft them?

Darrow was given the whole weekend, during which eight scientists dictated 60,000 words to stenographers.[15] Copies were given to the press; excerpts were read aloud in the courtroom. Far from being excluded, the testimony of Darrow's witnesses occupies 54 pages of the trial transcript. The decision to stay off the stand, and submit only written affidavits, was not made by a bigoted judge, but by the defense itself, in an effort to escape cross-examination. The ploy worked. Ironically, most of the evolutionary "evidence" Darrow's experts discussed—Piltdown Man, "useless" organs, embryonic recapitulation—now sits in the trash heap of discredited ideas.

IN THE MOVIE, Brady wins the case through a vicious betrayal that reveals much about the writers' view of Christian faith. The Scopes character, Bert Cates, is engaged to Rachel Brown, daughter of the town preacher. Reverend Brown tells Rachel she must leave Bert. Rachel refuses. He asks why. She says, "I love him, Pa." He says, "No, no, that is the love of Judas—this man has nothing to offer you but sin." Rachel asks: "Why do you hate him so?" Her father says, with a malicious expression: "Because I love God and I hate his enemies." (Got that, teens of the sixties? Only religion and parents stand in the way of "true love." Indeed, Cates tells Rachel: "It's his church or our house—you can't live in both.")

Reverend Brown orders Rachel to beg forgiveness. Refusing, she assertively confronts her father with how unloving he has been since her childhood. The reverend nearly has a nervous breakdown, falling on his knees and babbling Bible verses.

Later, Reverend Brown leads a prayer meeting which looks more like a storm trooper rally:

> Reverend Brown: Do we curse the man who denies the Word?
>
> Crowd (wild-eyed, frenzied): Yes!
>
> Brown: Do we call down hellfire on the man who has sinned against the Word?
>
> Crowd: Yes!
>
> Brown (looking toward heaven): O Lord of the Tempest and the Thunder! Strike down this sinner, as Thou didst Thine enemies of old, in the days of the Pharaohs! Let him know the terror of Thy sword! For all eternity, let his soul writhe in anguish and damnation!
>
> Rachel (sobbing): No, Pa! Don't pray to destroy Bert!
>
> Brown: Lord, we ask the same curse for those who ask grace for this sinner—though they be blood of my blood, and flesh of my flesh!

At this point, Matthew Harrison Brady intervenes, calling for moderation and forgiveness. He breaks up the prayer meeting, and walks Rachel home, apparently to comfort her. For a moment, the writers seem to be showing a more balanced picture of Brady.

However, we later learn it's a ruse to milk Rachel for information about Bert Cates! The next day, Brady calls her as a witness, insisting that she repeat to the courtroom what she said the night before. Rachel protests:

"Mr. Brady, I *confided* in you." But he and the judge compel her to reveal what Bert told her during their most intimate moments— personal thoughts showing he had some doubts about religion. Brady cruelly forces her to the point of tears, saying: "Tell it, tell it all, tell it, tell it, tell it!" In the movie, it is this testimony that convicts Cates.

That night, Rachel goes to Brady's rooming house. She tells his wife: "I turned to your husband for help. He encouraged me to open up my heart to him. And then he twisted my words. He tricked me! Why? Why did he do it?" She breaks down, sobbing, then tells Mrs. Brady: "If he could do such an evil thing, he must be an evil man. And everything he stands for must be evil too!" The next day, Rachel tells Bert: "I left my father." True love has won!

IN REAL LIFE, John Scopes had no girlfriend at the time of the trial, and no women were ever called to testify.

IN THE MOVIE, the judge cites Drummond with contempt of court because he makes an impassioned speech about truth and justice, denying that his client has received a fair trial (which in the movie, of course, he hasn't).

IN REAL LIFE, Darrow's contempt citation was for repeatedly insulting and interrupting the judge. The spark was Judge Raulston's ruling that Bryan could cross-examine defense experts if they took the stand. We have already quoted Darrow's testy remark that the purpose of cross-examination is to "show prejudice." Let's read excerpts from the ensuing exchange, which led to the citation:

> The Court—Courts are a mockery . . .
>
> Mr. Darrow—They are often that, your honor.
>
> The Court—. . . when they permit cross-examination for the purpose of creating prejudice.
>
> Mr. Darrow—I submit, your honor, there is no sort of question that they are not entitled to cross-examine, but all this evidence is to show what we expect to prove and nothing else, and can be nothing else.
>
> The Court—I will say this: If the defense wants to put their proof in the record, in the form of affidavits, of course they can do that. If they put the witness on the stand and the state desires to cross-examine them, I shall expect them to do so.
>
> Mr. Darrow—We except to it and take an exception.
>
> The Court—Yes sir; always expect this court to rule correctly.
>
> Mr. Darrow—No, sir, we do not.
>
> (Laughter) . . .
>
> The Court—I would not say . . .
>
> Mr. Darrow—If your honor takes half a day to write an opinion . . .

> The Court—I have not taken . . .
>
> Mr. Darrow—We want to make statements here of what we expect to prove. I do not understand why every request of the state and every suggestion of the prosecution should meet with an endless waste of time, and a bare suggestion of anything that is perfectly competent on our part should be immediately overruled.
>
> The Court—I hope you do not mean to reflect upon the court?
>
> [At this point, Darrow turned his back on the judge and hunched his shoulders.]
>
> Mr. Darrow—Well, your honor has the right to hope.
>
> [laughter][16]

Raulston did not charge Darrow with contempt in the heat of anger, as the movie judge does. He made the citation the following day, after reviewing the court transcript. And incidentally, as in the film, he dropped the citation as soon as Darrow apologized.

IN THE MOVIE, Cates and Drummond hang on pins and needles waiting for the jury's decision. When a "guilty" verdict is read, gloom falls on the defendant and his brave attorney. Brady and the prosecution exult. Bigotry and ignorance have won the day.

IN REAL LIFE, there was no suspense. On the last day of the trial, *Darrow himself* changed Scopes's plea to "guilty." Here are his words:

> Mr. Darrow—Let me suggest this. We have all been here quite a while and I say it in perfectly good faith, we have no witnesses to offer, no proof to offer on the issues that the court has laid down here, that Mr. Scopes did teach what the children said he taught, that man descended from a lower order of animals—we do not mean to contradict that, and I think to save time we will ask the court to bring in the jury and instruct the jury to find the defendant guilty.[17]

Yow! What goes on? Why did Darrow plead Scopes "not guilty," then do an about-face? The answer lies in his famous interrogation of William Jennings Bryan. Again, this is what the trial is primarily remembered for—that Darrow trounced Bryan, and that evolution thus trounced fundamentalism. Let's look at that famous debate.

IN THE MOVIE, Drummond, denied the right to quote Darwin or call scientific witnesses, is brooding in his hotel room. "What I need is a miracle," he says.

Hornbeck, the journalist, tosses him a Bible, saying: "Miracle? Here's a whole bagful. Courtesy Matthew Harrison Brady."

Drummond holds the Bible, thinking and smiling. An imaginary lightbulb pops above his head. Hm . . . If they won't let him ask science questions, he'll get 'em on the Bible. That egotist Brady would never pass up a chance to defend "the Good Book."

IN REAL LIFE, this decision was anything but spontaneous. A bitter critic of Christianity, Darrow had crafted most of his Bible questions years earlier. He had long yearned to debate Bryan. The night before the interrogation, he rehearsed it with Kitley Mather, one of his academic witnesses.[18]

Did Darrow win his confrontation with Bryan? Yes, but not nearly as convincingly as in the movie, and he succeeded for a plain reason. When a trial witness is interrogated, he may only answer the questions asked. Furthermore, he may not ask any questions himself. Thus Darrow totally controlled the exchange. He took the offensive throughout, while Bryan could only assume the *defensive*, answering questions, asking none. Normally, of course, no politician consents to debate under such one-sided conditions.

Why, then, in heaven's name, did Bryan? First, because Darrow baited him by publicly branding him a coward. Over the weekend, he told the press: "Bryan is willing to express his opinions on science and religion where his statements will not be questioned, but Bryan has not dared to test his views in open court under oath. . . ."[19]

When Darrow called Bryan as a witness that Monday, chief prosecutor Tom Stewart protested, but Bryan declared, to great applause: "I am simply trying to defend the word of God against the greatest atheist or agnostic in the United States. I want the papers to know I am not afraid to get on the stand and let him do his worst."[20]

But there was a more significant reason why Bryan agreed: believing that afterwards he'd have the opportunity to *question Darrow on evolution.* This was important to Bryan since he had been denied cross-examination of Darrow's experts.

> Mr. Bryan—If your honor please, I insist that Mr. Darrow can be put on the stand, and [defense attorneys] Mr. Malone and Mr. Hays.
>
> The Court—Call anybody you desire. Ask them any questions you wish.[21]

Darrow strung Bryan along, letting him believe this would happen:

> The Witness [Bryan]: I want him [Darrow] to have all the latitude he wants. For I am going to have some latitude when he gets through.
>
> Mr. Darrow—You can have latitude and longitude.[22]

But as we will see, Darrow apparently had no intention of going on the stand! Now let's inspect the interrogation.

IN THE MOVIE, Brady is a Bible literalist:

> Drummond: You believe that every word written in this book should be taken literally?
>
> Brady: Everything in the Bible should be accepted, exactly as it is given there.

IN REAL LIFE, we find Lawrence and Lee lifted Bryan's answer out of context:

> Darrow: Do you claim that everything in the Bible should be literally interpreted?
>
> Bryan: I believe everything in the Bible should be accepted as it is given there; some of the Bible is given illustratively. For instance: "Ye are the salt of the earth." I would not insist that man was actually salt, or that he had flesh of salt, but it is used in the sense of salt as saving God's people.[23]

IN THE MOVIE, Drummond asks Brady about sex:

> Drummond: You're up here as an expert on the Bible. What is the Biblical evaluation of sex?
>
> Brady: It is considered "Original Sin."

Lawrence and Lee thus established Brady as a prude.

IN REAL LIFE? Darrow never asked Bryan about sex. Incidentally, the Bible says *adultery*, not sex, is sinful.

IN THE MOVIE, Drummond asks Brady how old the Earth is:

> Brady: A fine Biblical scholar, Bishop Ussher, has determined for us the exact date and hour of the Creation. It occurred in the year 4,004 B.C.
>
> Drummond: Well, uh, that's Bishop Ussher's opinion.
>
> Brady: It is not an opinion. It is a literal fact, which the good Bishop arrived at through careful computation of the ages of the prophets as set down in the Old Testament. In fact, he determined that the Lord began the Creation on the 23rd of October, 4,004 B.C. at—uh, at 9 A.M.
>
> Drummond: That Eastern Standard Time?

IN REAL LIFE, here's what was said:

> Q—Mr. Bryan, could you tell me how old the earth is?
>
> A—No, sir, I couldn't.
>
> Q—Could you come anywhere near it?
>
> A—I wouldn't attempt to. I could possibly come as near as the scientists do, but I had rather be more accurate before I give a guess.[24]

IN THE MOVIE, Brady crumbles as Drummond brings his interrogation to a climax:

Brady: It is the revealed Word of the Almighty God spake to the men who wrote the Bible.

Drummond: How do you know that God didn't "spake" to Charles Darwin?

Brady: I know, because God tells me to oppose the evil teachings of that man.

Drummond: Oh. God speaks to you.

Brady: Yes.

Drummond: He tells you what is right and wrong.

Brady: Yes.

Drummond: And you act accordingly?

Brady: Yes.

Drummond: So you, Matthew Harrison Brady, through oratory, legislature, or whatever, you pass on God's orders to the rest of the world! Well, meet the "Prophet From Nebraska"!

Brady begins cracking up. Finally—even after being dismissed as a witness—all he can do is frantically shout the names of the books of the Bible. The fundamentalists in the courtroom are visibly disillusioned and even angry with their hero.

IN REAL LIFE, nothing remotely resembling this sequence occurred.

Sure, Darrow scored some points. One of his wittiest moments came while pursuing Bryan on the date of the Flood:

Q—What do you think?

A—I do not think about things I don't think about.

Q—Do you think about things you *do* think about?[25]

This resulted in an outburst of courtroom laughter. And you can be certain that, while Lawrence and Lee invented most of their dialogue, they kept *this*. After all, we must have some reality, mustn't we?

On the other hand, the playwrights took care to eliminate Darrow's surly remarks, such as his reference to Christianity as "your fool religion."[26]

The truth is, Bryan often gave as good as he got. One Darrow strategy was to list various esoteric subjects, such as philology, and ask Bryan if he had ever studied them. Since Bryan was forced to keep answering "No," it made him appear ignorant. However, Bryan soon discerned that Darrow did not necessarily know the answers to his own questions:

Q—Do you know about how many people there were on this earth 3,000 years ago?

A—No.

Q—Did you ever try to find out?

A—When you display my ignorance, could you not give me the facts, so I would not be ignorant any longer? Can you tell me how many people there were when Christ was born?

Q—You know, some of us might get the facts and still be ignorant.

A—Will you please give me that? You ought not to ask me a question when you don't know the answer to it.

Q—I can make an estimate.

A—What is your estimate?

Q—Wait until you get to me.[27]

Here we see Darrow still baiting Bryan with the promise that their roles would soon be reversed.

But the next morning, Bryan sat stunned as Darrow changed Scopes's plea from "not guilty" to "guilty," thus ending the trial. The judge gave Darrow an ostensible excuse by saying he planned to expunge the chaotic Bryan-Darrow interrogation from the record. But it is unlikely that Darrow ever planned to take the stand. It is well established that he intended to keep the eloquent Bryan from making a closing statement. As Darrow biographer Kevin Tierney noted:

> Darrow, realizing that Bryan might make a comeback by giving a final address to the jury, pleaded Scopes guilty and waived the defense's right to a closing speech, thereby under Tennessee law depriving the prosecution of the chance to address the court.[28]

Darrow himself wrote in his autobiography:

> I made a complete and aggressive opening of the case. I did this for the reason that we never at any stage intended to make any [closing] arguments in the case. . . . By not making a closing argument on our side we could cut him [Bryan] out.[29]

The trial had never been about John Scopes's guilt or innocence. Its purpose had been to disseminate Darwinism and assail fundamentalism. Darrow accomplished both. He had gotten his witnesses' testimonies into the record without their being cross-examined; and he had roughed up Bryan on the Bible, then prevented Bryan from reciprocating. Give Darrow credit—he was a great tactician.

Williams Jennings Bryan thought it was a football game. He let Darrow go on offense first; Darrow drove downfield and, after a hard battle, scored a touchdown. Now, as Bryan stood awaiting a return kickoff, Darrow announced that the game was over. He proclaimed himself winner and was carried off the field on the media's shoulders.

Bryan cynically commented that day: "... I think it is hardly fair for them to bring into the limelight my views on religion and stand behind a dark lantern that throws light on other people, but conceals themselves."[30]

IN THE MOVIE, when the jury convicts Bert Cates, the judge decides to be lenient, and fines him only $100. Brady is wildly upset.

> Brady: Did your honor say one hundred dollars?
>
> Judge: That is correct. That seems to conclude the business of the trial . . .
>
> Brady: Your honor, the prosecution takes exception! Why, the issues are so titanic, the court must mete out more drastic punishment!

IN REAL LIFE, Bryan had opposed having *any* penalties attached to Tennessee's Butler Act.[31] Regarding the Scopes case, he said:

> I don't think we should insist on more [than] the minimum fine, and I will let the defendant have the money to pay if he needs it.[32]

Scopes *was* fined $100, the minimum under the law, but was never required to pay; the Tennessee Supreme Court later disallowed it on a technicality.

IN THE MOVIE, Bert Cates's career is over. Facing the judge for sentencing, he says: "I do not have the eloquence of some of the men you have heard in the last few days. I'm just a schoolteacher." A woman in the courtroom shouts: "Not any more you ain't!" Cates says sheepishly: "I *was* a schoolteacher."

IN REAL LIFE, Scopes wrote in his autobiography: "I could have continued teaching in Dayton. Doc Robinson, as president of the school board, offered me my old job of coaching and teaching math and physics. . . ."[33] But Scopes opted instead to undertake graduate studies.

IN THE MOVIE, after his showdown with Drummond, Brady is obsessed with a speech he wants to deliver in court. He says to his wife in their hotel room: "My speech! Where's my speech? I'll make them listen! Where's my speech? I must have it!"

He then sobs to his wife: "Mother! They laughed at me!" Mrs. Brady holds her husband like a child and says, "Hush, baby."

> Brady: I can't stand it when they laugh at me!
>
> Mrs. Brady: It's all right, baby. It's all right.

The next day, after the trial concludes, Brady attempts to make his "speech" to the courtroom. But no one is interested—reporters and hawkers are busy talking. Brady pathetically babbles religious phrases, trying to shout above the crowd noise. Drummond, Cates and Hornbeck watch him with looks of disgust bordering on pity. Even his wife is appalled. Then Brady falls down with a big "thud" and dies. Having assassinated the man's character throughout the script, the writers now kill him off for real, like, "There! Take that, ya lousy stinkin' bigot!" To add to a touch of sadistic humor, Brady had been waving a fan bearing the name of a funeral parlor.

IN REAL LIFE, spectators laughed at Darrow as much as Bryan. And what was this "speech" Bryan was supposedly obsessed with? He had prepared a *closing statement*; every attorney does in a jury case. But as we have seen, Darrow, knowing Bryan was a powerful orator, nixed that too by changing Scopes's plea to guilty.

Bryan made no attempt to deliver a lengthy speech that day in court, and certainly didn't die there. He did die of a stroke five days later, but as he pursued a very vigorous schedule over those days, he was clearly not the "broken man" some have claimed. Bryan was elderly, had a bad heart and diabetes, and was also nicked by a passing automobile after the trial. Doubtless the case's rigors took their toll as well.

The movie's smear of Bryan knew few limits. He was known to have a big appetite, a condition his diabetes probably exacerbated. But the film goes to absurd proportions, with Brady gorging himself on fried chicken right in the courtroom. Brady is not only a liar, but a moron, always disarmed by the wit of E. K. Hornbeck, the journalist based on H. L. Mencken. By contrast, Drummond is brilliant, kind, courageous, honest, and even gets along with Brady's wife much better than Brady!

The people of Tennessee—except Cates, his students and girlfriend—are also denigrated as ignorant bigots. In a masterful stroke of subtlety, however, Lawrence and Lee did *not* begin every sentence spoken by a Tennessean with "Duh."

IN THE MOVIE, when E. K. Hornbeck learns of Brady's death, he says he "died of a busted belly." Drummond chides Hornbeck for being so callous. The audience sees the attorney is thus gracious to his enemies.

IN REAL LIFE, it was Darrow *himself* who said Bryan "died of a busted belly." (Mencken reportedly said: "We killed the son-of-a-bitch!")[34]

Lest anyone think me alone in this assessment of *Inherit the Wind*, I quote *Time* magazine's reaction: "The script wildly and unjustly caricatures the fundamentalists as vicious and narrow-minded hypocrites, just as wildly and unwisely idealizes their opponents, as personified by Darrow."[35] Critic Andrew Sarris called it "bigotry in reverse."[36] The *New Yorker* commented that "history has not been increased but almost fatally diminished. . . . the picturing of Dayton as a community composed entirely of backwoods religious maniacs, which apparently wasn't the case at all, makes the play a much too elementary study in black and white."[37]

Constitutional scholar Gerald Gunther said it was the only play he ever walked out on:

> I ended up actually sympathizing with Bryan, even though I was and continue to be opposed to his ideas in the case, simply because the playwrights had drawn the character in such comic strip terms.[38]

In the movie's theatrical trailer, after showing some clips of the Drummond-Brady debate, producer-director Stanley Kramer told audiences: "The winner? You'll have to make that decision for yourselves when you see *Inherit the Wind.*"

Oh, thanks, Stan! We can decide for ourselves! We can choose Brady, who's gluttonous, hypocritical, ignorant, mean-spirited, prudish, and laughs at his own bad jokes—or choose Drummond, who's witty, courageous, generous, sincere, and broad-minded.

Inherit the Wind asserts that John Scopes was convicted because he didn't receive a fair trial. It falsely claims that all evidence supporting his case was disallowed.

Ironically, that's what *Inherit* did. It prosecuted William Jennings Bryan, the people of Tennessee, and Christians by showing the "jury" (the audience) only one side of the story—a fabricated one.

When the fictitious "Reverend Brown" whipped glassy-eyed fundamentalists into a frenzy—helping turn them into a lynch mob—audiences were expected to be appalled at how the minister's propagandizing built hatred into his followers. Ironically, it was the film's viewers themselves, watching this scene, who were being propagandized and encouraged to hate.

Throughout the movie, defense attorney Henry Drummond speaks out against "bigotry, ignorance and hate." Those words pretty well summarize *Inherit the Wind.*

PLATE 62. Clarence Darrow

PLATE 63. John T. Scopes

PLATE 64. William Jennings Bryan

PLATE 65. Prosecutor Tom Stewart went on to become a U.S. Senator.

PLATE 66. Scopes and Dayton citizens reenact the scene at Doc Robinson's drugstore where the teacher was recruited to say he broke the law. Scopes is seated, second from left. George Rappleyea is looking over his shoulder.

PLATE 67. Clarence Darrow (extreme right) shakes hands with Judge John T. Raulston after the latter dismissed his contempt citation.

PLATE 68. In the movie, Henry Drummond (Spencer Tracy) delivers one knockout punch after another when he cross-examines Matthew Harrison Brady (Frederic March).

PLATE 69. Real life, July 20, 1925: Clarence Darrow (right) cross-examining William Jennings Bryan. Due to the heat and the size of the crowd, Judge Raulston held court outside that day. No knockout punches were landed.

PLATE 70. Fictitious Reverend Brown goes bonkers while praying. Actor Claude Akins went on to star in Ryder Truck commercials.

CHAPTER 18

Have You Murdered Anybody Since Breakfast?

If Darwinism was simply a hypothesis, whose greatest impact was stimulating scholarly debates in *Scientific American*, then it wouldn't trouble us. Unfortunately, evolution has had severely negative social impact.

While Darwin did not invent racism, he did provide "scientific" grounds for it. As John C. Burnham noted in *Science*:

> What was new in the Victorian period was Darwinism. . . . Before 1859, many scientists had questioned whether blacks were of the same species as whites. After 1859, the evolutionary schema raised additional questions, particularly whether or not Afro-Americans could survive competition with their white near-relations. The momentous answer was a resounding no. . . . The African was inferior because he represented the "missing link" between ape and Teuton. . . .[1]

The subtitle of *The Origin of Species* was *The Preservation of Favoured Races in the Struggle for Life*. Although Darwin penned that in an animal context, extending it to human races was a small leap of logic. He would later write:

> I could show fight on natural selection having done and doing more for the progress of civilization than you seem inclined to admit. . . . The more civilized so-called Caucasian races have beaten the Turkish hollow in the struggle for existence. Looking to the world at no very distant date, what an endless number of the lower races will have been eliminated by the higher civilized races throughout the world.[2]

Racism was almost universal among leading early evolutionists, many of whom believed the races had evolved separately. Thomas Huxley, "Darwin's bulldog," wrote:

> It may be quite true that some negroes are better than some white men; but no rational man, cognizant of the facts, believes that the average Negro is the equal, still less the superior, of the white man. And if this be true, it is simply incredible that, when all his disabilities are removed, and our prognathous relative has a fair field and no favour, as well as no oppressor, he will be able to compete successfully with his bigger-brained and smaller-jawed rival, in a contest which is to be carried out by thoughts and not by bites. The highest places within the hierarchy of civilization will assuredly not be within the reach of our dusky cousins. . . .[3]

Ernst Haeckel, the great popularizer of Darwinism in Germany, was even more severe:

> The mental life of savages rises little above that of the higher mammals, especially the apes, with which they are genealogically connected. . . . Their intelligence moves within the narrowest bounds, and one can no more (or no less) speak of their reason than of that of the more intelligent animals. . . . These lower races (such as the Veddahs or Australian negroes) are psychologically nearer to the mammals (apes or dogs) than to civilized Europeans; we must, therefore, assign a totally different value to their lives.[4]

Henry Osborn, director of the American Museum of Natural History, was the evolutionist who imagined an entire civilization—"Nebraska Man"—based on a single pig's tooth. He was also one of William Jennings Bryan's most outspoken critics in the 1920s. Osborn declared:

> The standard of intelligence of the average adult Negro is similar to that of the eleven-year-old youth of the species *Homo Sapiens*.[5]

Thus, like most Darwinists of his day, Osborn classed blacks as an inferior subdivision of humanity. He said:

> If an unbiased zoologist were to descend upon the earth from Mars and study the races of man with the same impartiality as the races of fishes, birds, and mammals, he would undoubtedly divide the existing races of man into several genera and into a very large number of species and subspecies.[6]

Edwin G. Conklin, professor of biology at Princeton, wrote in his book *The Direction of Human Evolution* (1923):

> Comparison of any modern race with the Neanderthal or Heidelberg types shows that all have changed, but probably the negroid races more closely resemble the original stock than the white or yellow races. . . . The greatest danger which faces any superior race is that of amalgamation with inferior stock and the consequent lowering of inherited capacities. . . . Every consideration should lead those who believe in the superiority of the white race to strive to preserve its purity and to establish and maintain the segregation of the races. . . .[7]

Even Hunter's *Civic Biology*—the book John Scopes allegedly taught to his students—said in its discussion of races that Caucasians were "the highest type of all."[8]

As we have noted, embryonic recapitulation, the theory propounded by Ernst Haeckel, theorized that the embryo not only moved through the animal phases of its evolutionary history, but racial phases as well. Stephen Jay Gould wrote: "Haeckel and his colleagues also invoked recapitulation to affirm the racial superiority of northern European whites."[9]

These ideas led to eugenics—the campaign to improve humanity through selective breeding. In Britain, Charles Darwin's son Leonard became president of the Eugenics Education Society. In the U.S., the movement caught fire in the early twentieth century. By 1935, 35 states had enacted laws requiring the sexual isolation and sterilization of "unfit" people—including the retarded, the "feeble-minded," chronic criminals, and even epileptics. Proposed legislation targeted tuberculosis sufferers, alcoholics, the blind and the homeless. About 70,000 Americans were involuntarily sterilized before the practice was stopped.[10]

But it was Germany, of course, that brought these ideas to their fullest fruition. In his 1904 book *The Wonders of Life*, Haeckel wrote:

> Hence the destruction of abnormal new-born infants—as the Spartans practiced it, for instance, in selecting the bravest—cannot rationally be classed as "murder," as is done in even modern legal works. We ought rather to look upon it as a practice of advantage both to the infants destroyed and to the community. . . .
>
> The ancient Spartans owed a good deal of their famous bravery, their bodily strength and beauty, as well as their mental energy and capacity, to the old custom of doing away with new-born children who were born weakly or crippled. . . . When I pointed out the advantages of this Spartan selection for the improvement of the race in 1868 (chapter vii. of the *History of Creation*) there was a storm of pious indignation in the religious journals, as always happens when pure reason ventures to oppose the current prejudices and traditional beliefs. But I ask: What good does it do to humanity to maintain artificially and rear the thousands of cripples, deaf-

mutes, idiots, etc., who are born every year with an hereditary burden of incurable disease?[11]

Haeckel advocated doing away with undesirable adults as well:

> Hundreds of thousands of incurables—lunatics, lepers, people with cancer, etc.—are artificially kept alive in our modern communities, and their sufferings are carefully prolonged, without the slightest profit to themselves or the general body. We have strong proof for this in the statistics of lunacy and the growth of asylums and nerve-sanatoria. . . . What an enormous mass of suffering these figures indicate for the invalids themselves, and what a vast amount of trouble and sorrow for their families, what a huge private and public expenditure! How much of this pain and expense could be spared if people could make up their minds to free the incurable from their indescribable torments by a dose of morphia![12]

Evolution strongly influenced German philosopher Friedrich Nietzsche. Calling Darwin one of the three greatest men of his century (he put Napoleon first), he denounced Christianity and declared: "God is dead." Nietzsche, of course, advanced the idea of the "superman" and "master race." He extolled ancient warriors and disparaged the weak and humble.

William Jennings Bryan aptly observed in 1925 that "the military textbooks in due time gave Germany the doctrine of the superman translated into the national policy of the superstate aiming at world power."[13] Later, Hitler and the Nazis took the ideas of Haeckel and Nietzsche to their zenith. As philosopher Gertrude Himmelfarb remarked in 1959:

> From the "preservation of favoured races in the struggle for life" [Darwin's subtitle to *Origin of Species*], it was a short step to the preservation of favoured individuals, classes, or nations—and from their preservation to their glorification. Social Darwinism has often been understood in this sense: as a philosophy, exalting competition, power, and violence over convention, ethics, and religion. Thus it has become a portmanteau of nationalism, imperialism, militarism and dictatorship, of the cults of the hero, the superman, and the master race. . . . Recent expressions of this philosophy, such as *Mein Kampf*, are, unhappily, too familiar to require exposition here. And it is by an obvious process of analogy and deduction that they are said to derive from Darwinism. Nietzsche predicted that this would be the consequence if the Darwinian theory gained general acceptance. . . .[14]

Hitler sought to achieve "racial purity" through elimination of Jews and perpetuation of an Aryan "master race." At least two million people were forcibly sterilized in Nazi Germany,[15] to say nothing of the millions who perished in death camps. In his demented way, Hitler was fulfilling this prediction Darwin made in *The Descent of Man*:

> At some future period, not very distant as measured by centuries, the civilized races will almost certainly exterminate, and replace, the savage races throughout the world. . . . The break between man and his nearest allies will then be wider, for it will intervene between man in a more civilized state, as we may hope, even than the Caucasian, and some ape as low as a baboon, instead of as now between the negro or Australian and the gorilla.[16]

Darwin, in that same book, had said:

> With savages, the weak in body or mind are soon eliminated; and those that survive commonly exhibit a vigorous state of health. We civilized men, on the other hand, do our utmost to check the process of elimination: we build asylums for the imbecile, the maimed, and the sick; we institute poor-laws; and our medical men exert their utmost skill to save the life of every one to the last moment. . . . Thus the weak members of civilized societies propagate their kind. No one who has attended to the breeding of domestic animals will doubt that this must be highly injurious to the race of man. . . . excepting in the case of man himself, hardly anyone is so ignorant as to allow his worst animals to breed.[17]

Jerry Bergman of Northwest Technical College notes:

> A review of the writings of Hitler and contemporary German biologists finds that Darwin's theory and writings had a major influence upon Nazi policies. Hitler believed that the human gene pool could be improved by selective breeding, using the same techniques that farmers used to breed a superior strain of cattle. In the formulation of his racial policies, he relied heavily upon the Darwinian evolution model, especially the elaborations by Spencer and Haeckel.[18]

Sir Arthur Keith, president of the British Association for the Advancement of Science, wrote in the 1940s:

> The German Fuhrer, as I have consistently maintained, is an evolutionist; he has consciously sought to make the practice of Germany conform to the theory of evolution.[19]

Keith further observed:

> We see Hitler devoutly convinced that evolution produces the only basis for a national policy. . . . The means he adopted to secure the destiny of his race and people were organized slaughter, which has drenched Europe in blood. . . . Such conduct is highly immoral as measured by every scale of ethics, yet Germany justifies it; it is consonant with tribal or

> evolutionary morality. Germany has reverted to the tribal past, and is demonstrating to the world, in their naked ferocity, the methods of evolution, with this difference—what were mere border forays between tribes have become the clash of massed millions using the forked lightning of modern science.[20]

Hitler delineated his views in his book *Mein Kampf*, which means "My Struggle." Darwin, of course, had defined life as a "struggle for existence," and Hitler largely saw the struggle as between races. George Stein, in his *American Scientist* article "Biological Science and Nazism," noted:

> The Germans, who focused on selection and the "struggle," or *Kampf* as it was translated, were closer to the radical insight of Darwin's efforts.[21]

Hitler's worldview clearly reflected Darwin's belief that higher states of evolution are achieved when the "fittest" survive the struggle for existence. In *Mein Kampf*, he said of races:

> The stronger must dominate and not blend with the weaker, thus sacrificing his own greatness. Only the born weakling can view this as cruel, but he after all is only a weak and limited man; for if this law did not prevail, any conceivable higher development of organic living beings would be unthinkable.[22]

As noted German philosopher Erich Fromm said:

> The "religion" of social Darwinism belongs to the most dangerous elements within the thoughts of the last century. It aids the propagation of ruthless national and racial egoism by establishing it as a moral norm. If Hitler believed in anything at all, then it was in the laws of evolution which justified and sanctified his actions and especially his cruelties.[23]

Something for everyone

If Darwin contributed to the development of Naziism, he did no less for its totalitarian cousin, communism. Having once been a leftist atheist myself, I'm not hesitant to say that evolution and Marxism go hand in hand. Marx denounced religion as "the opium of the people," and in nearly every nation where Communists took power, the church was, if not abolished outright, neutralized in effect.

George Stein noted in *American Scientist*: "Marx himself viewed Darwin's work as confirmation by the natural sciences of his own views. . . ."[24] There were many extremists before Darwin published *Origin*. But since religious faith prevailed among most of the world's leading scientists, it was hard persuading the masses to accept radical ideologies. Darwin, however, opened the door by providing a "scientific" rationale for denying God.

While Hitler envisioned the "struggle for existence" as between races, Marx saw it between classes. He said: "Darwin's book is very important and serves me as a basis in natural science for the class struggle in history."[25] Marx sent the naturalist proof-sheets of *Das Kapital*, and offered to dedicate it to him, but Darwin politely declined, noting "the concern it might cause some members of my family, if in any way I lent my support to direct attacks on religion."[26]

Soviet dictator Joseph Stalin murdered millions. Like Darwin, he began as a theology student. And like Darwin, evolution transformed his life. In 1940, a book was published in Moscow entitled *Landmarks in the Life of Stalin* by Emelian Yaroslavsky. In it we read:

> At a very early age, while still a pupil in the ecclesiastical school, Comrade Stalin developed a critical mind and revolutionary sentiments. He began to read Darwin and became an atheist.
>
> G. Glurdjidze, a boyhood friend of Stalin's, relates:
>
> "I began to speak of God. Joseph heard me out, and after a moment's silence, said:
>
> "'You know, they are fooling us, there is no God. . . .'
>
> "I was astonished at these words. I had never heard anything like it before.
>
> "'How can you say such things, Soso?' I exclaimed.
>
> "'I'll lend you a book to read; it will show you that the world and all living things are quite different from what you imagine, and all this talk about God is sheer nonsense,' Joseph said.
>
> "'What book is that?' I enquired.
>
> "'Darwin. You must read it,' Joseph impressed on me."[27]

Stein noted that "even Mao Tse-tung regarded Darwin, as presented by the German Darwinists, as the foundation of Chinese scientific socialism."[28] Mao was yet another who decimated millions.

And why not? Darwin had "proven" that men were not God's creation. Instead, they were descended from bacteria, fish and lizards. So in the minds of Hitler, Stalin and Mao, why not *treat* people as animals? Why not herd them like cattle into boxcars bound for concentration camps and gulags?

Darwinism also helped justify the ruthless practices of monopolists such as Andrew Carnegie and John D. Rockefeller. To them, the "struggle for existence" meant destroying business competition; if they succeeded, this was merely "survival of the fittest." As Carnegie said:

> While the law may sometimes be hard for the individual, it is best for the race, because it ensures the survival of the fittest in every department.[29]

Carnegie had once believed in Christianity, but abandoned it for Darwinism. He wrote:

> When I, along with three or four of my boon companions, was in this stage of doubt about theology, including the supernatural element, and indeed the whole scheme of salvation through vicarious atonement and all the fabric built upon it, I came fortunately upon Darwin's and Spencer's works. . . . I remember that light came as in a flood and all was clear. Not only had I got rid of theology and the supernatural, but I had found the truth of evolution. "All is well since all grows better" became my motto, my true source of comfort.[30]

John D. Rockefeller reputedly said, "The growth of a large business is merely a survival of the fittest."[31] The Rockefellers, while maintaining Christian trappings, financed the ministry of Harry Emerson Fosdick, brother of one of their attorneys. Fosdick preached "modern" Christianity on radio for twenty years, embracing evolution and downgrading the Bible's early books to mythology.[32] During the Scopes trial, when a philanthropist pledged $10,000 to help found a university named after William Jennings Bryan, John D. Rockefeller, Jr., retaliated the very same day with a $1 million donation to the liberal University of Chicago Divinity School.[33]

Evolution inspiring both communists and archcapitalists? Not as surprising as it sounds. Both stubbornly oppose the values of Biblical Christianity. They are simply on different ends of the "class struggle."

Darwin wouldn't score well with today's women, either. He considered men more highly evolved; women, he claimed, had traits that "are characteristic of the lower races, and therefore of a past and lower state of civilization."[34] He said:

> The chief distinction in the intellectual powers of the two sexes is shown by man's attaining to a higher eminence, in whatever he takes up, than can woman—whether requiring deep thought, reason, or imagination, or merely the use of the senses and hands. . . . Thus man has ultimately become superior to woman. It is, indeed, fortunate that the law of the equal transmission of characters to both sexes prevails with mammals; otherwise it is probable that man would have become as superior in mental development to woman, as the peacock is in ornamental plumage to the peahen.[35]

Evolutionist Gustave Le Bon, a contemporary of Darwin, wrote:

> In the most intelligent races, as among the Parisians, there are a large number of women whose brains are closer in size to gorillas than to the

> most developed male brains. This inferiority is so obvious that no one can contest it for a moment; only its degree is worth discussion. All psychologists who have studied the intelligence of women, as well as poets and novelists, recognize today that they represent the most inferior forms of human evolution and that they are closer to children and savages than to an adult, civilized man. They excel in fickleness, inconstancy, absence of thought and logic, and incapacity to reason. Without a doubt there exist some distinguished women, very superior to the average man, but they are as exceptional as the birth of any monstrosity, as, for example, of a gorilla with two heads; consequently, we may neglect them entirely.[36]

Of course, evolution was not solely responsible for every ideology it helped foster. The genocides of the Nazis and communists would have revolted Darwin. Nor am I suggesting that today's evolutionists advocate totalitarianism or racism. In fact, Stephen Jay Gould has been a leader in exposing the racism of early Darwinists. However, evolution's deadly influences cannot be denied.

There is a courtroom scene in *Inherit the Wind* that runs as follows. Henry Drummond (Clarence Darrow) is questioning one of the students of Bert Cates (John Scopes):

> Drummond: Let's put it this way, Howard. All this fuss and feathers about evolution, do you think it hurt you any?
>
> Howard: Sir?
>
> Drummond: Did it do you any harm? You still feel reasonably fit? What Mr. Cates told you, did it hurt your baseball game any? Affect your pitching arm? [playfully punches the boy's right arm]
>
> Howard: No, sir. I'm a lefty.
>
> Drummond: A southpaw, huh? Still honor your mother and father?
>
> Howard: Sure.
>
> Drummond: Haven't murdered anybody since breakfast, have you?

I wish Joseph Stalin had been on the stand. He could have answered, quite truthfully, "Why, yes—several thousand people, in fact." But of course, why get concerned about a lot of "fuss and feathers"?

PLATE 71. Friedrich Nietzsche —ideological middleman between Darwin and Hitler

PLATE 72. Henry Fairfield Osborn viewed races as subspecies.

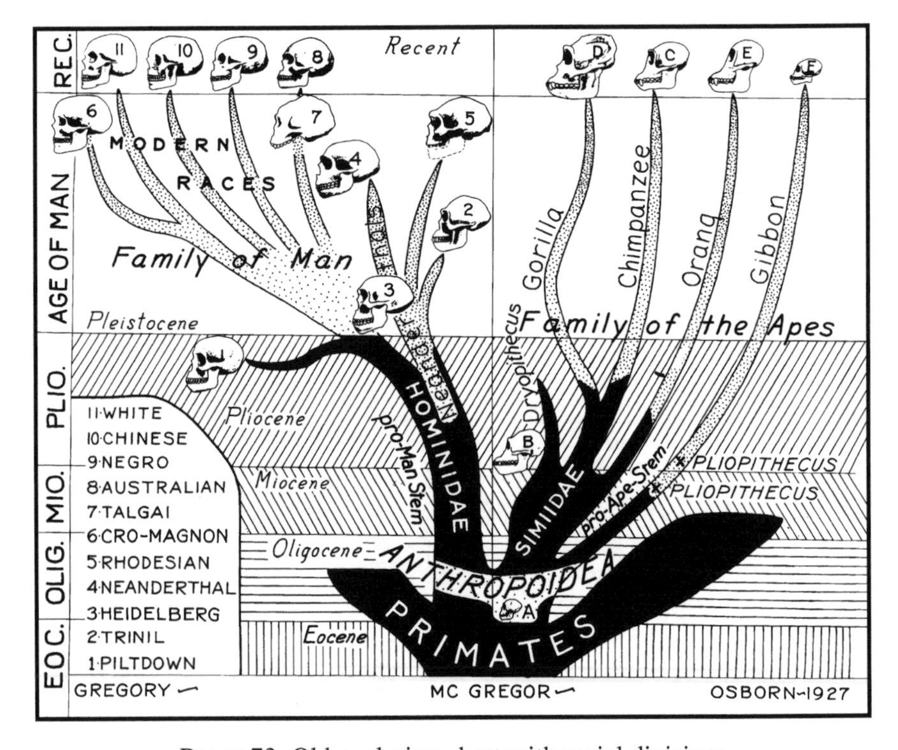

PLATE 73. Old evolution chart with racial divisions

CHAPTER 19

The Boomers Doomed

Shortly before the Scopes affair, Clarence Darrow had defended two murderers—Leopold and Loeb—in one of the twentieth century's most sensational trials. Leopold and Loeb were college students who murdered a boy just for thrills—to see if they could get away with it. Darrow pleaded for clemency, arguing that Leopold was not responsible for his actions, but had been driven to murder by reading Friedrich Nietzsche's works. During the Scopes trial, a chagrined Darrow listened as William Jennings Bryan quoted his defense of Leopold and Loeb:

> I will guarantee that you can go down to the University of Chicago today—into its big library and find over 1,000 volumes of Nietzsche, and I am sure I speak moderately. If this boy is to blame for this, where did he get it? Is there any blame attached because somebody took Nietzsche's philosophy seriously and fashioned his life on it? And there is no question in this case but what is true. Then who is to blame? The university would be more to blame than he is. The scholars of the world would be more to blame than he is. The publishers of the world—and Nietzsche's books are published by one of the biggest publishers in the world—are more to blame than he. Your honor, it is hardly fair to hang a 19-year-old boy for the philosophy that was taught him at the university.[1]

Who inspired Nietzsche? To a great extent, Darwin. Yet now, at the Scopes trial, Darrow argued that reading Darwin *couldn't possibly affect students' behavior*. You can be sure this insight of Bryan's was edited out of *Inherit the Wind*.

Some areas the last chapter discussed—Naziism, communism and monopolism—may seem remote and irrelevant to today's readers. So let's go closer to home: America's baby boomers.

I started this book by recalling how my sixth-grade teacher, Mr. Halpern, told our class that Earth was five billion years old and the Bible was wrong. No, we didn't go out and shoot anyone that day. But as thousands of Mr.

Halperns around the USA taught the same thing, they planted some pretty destructive seeds.

Unlike Darrow, William Jennings Bryan accurately foresaw the results if evolution were taught to children as fact: "If they believe it," he said, "they go back to scoff at the religion of their parents!"[2] He reasoned: If evolution turned *Darwin* from faith, would it do less to others?

And isn't that what happened? I quote Harvard professor E. O. Wilson, chief spokesperson for evolutionary sociobiology, and a bitter critic of Biblical Christianity:

> As were many persons from Alabama, I was a born-again Christian. When I was fifteen, I entered the Southern Baptist Church with great fervor and interest in the fundamentalist religion; I left at seventeen when I got to the University of Alabama and heard about evolutionary theory.[3]

That nicely sums what happened to us baby boomers. I wasn't raised religiously myself, but once sold on the "fact" of evolution, I became a dedicated atheist—faith stood no chance with me.

There was a reason why my generation bought evolution so easily. If you were young, you weren't hot about Biblical morality to begin with. But here was *teacher*, the same person who taught you math equations, saying—these weren't the words, but how we *translated* them—"Pssst! Hey, kids! You know that God stuff you've been hearing about? All hooey! There ain't no God, and that's a scientific fact, Jack! The Ten Commandments? They're bull! Forget 'em! Make up your *own* rules!" For rebellious teenagers, that message wasn't too hard to take. Suddenly, we had an excuse to do what we already wanted to do. Like, it was OK to get wasted.

The Bible had been Western civilization's central guiding document for nearly two thousand years. It defined moral values in absolutes. But now right and wrong were ours to decide. As activist Jeremy Rifkin put it in his book *Algeny*:

> We no longer feel ourselves to be guests in someone else's home and therefore obliged to make our behavior conform with a set of pre-existing cosmic rules. It is our creation now. We make the rules. We establish the parameters of reality. We create the world, and because we do, we no longer feel beholden to outside forces. We no longer have to justify our behavior, for we are now the architects of the universe. We are responsible to nothing outside ourselves, for we are the kingdom, the power, and the glory forever and ever.[4]

Thomas Huxley's grandson, Julian Huxley, was an atheist and one of evolution's foremost spokesmen in the twentieth century. He stated that

"Darwinism removed the whole idea of God as the Creator of organisms from the sphere of rational discussion."[5] But did we stop believing in God because evolution was proven—or did we accept evolution in order to deny God? As Dr. Michael Walker, senior lecturer in anthropology at Sydney University, has noted:

> One is forced to conclude that many scientists and technologists pay lip-service to Darwinian Theory only because it supposedly excludes a Creator from yet another area of material phenomena, and not because it has been paradigmatic in establishing the canons of research in the life sciences and the earth sciences.[6]

Julian's brother, author Aldous Huxley, an early advocate of the drug culture and sexual permissiveness, put it bluntly:

> I had motives for not wanting the world to have meaning; consequently assumed it had none, and was able without any difficulty to find satisfying reasons for this assumption. . . . For myself as, no doubt, for most of my contemporaries, the philosophy of meaninglessness was essentially an instrument of liberation. The liberation we desired was simultaneously liberation from a certain political and economic system and liberation from a certain system of morality. We objected to the morality because it interfered with our sexual freedom; we objected to the political and economic system because it was unjust.[7]

Unfortunately, penalties go with denying God. Teen suicides have reached tragic levels. There's an epidemic of "low self-esteem." Not surprising. One hundred years ago, students learned they were made in God's image. Today they are told, in effect: "The universe was created by chance. You, therefore, are here by accident. Your life has no purpose or meaning. There is no afterlife; you have no soul. You are just a blob of molecules, the result of horrid mutations, which turned out OK thanks to your animal ancestors winning a bloody tooth-and-claw struggle for existence. Your great-great grandparents, many times removed, were reptiles, their ancestors were fish, but ultimately you're descended from bacteria." I can't imagine why that would diminish anyone's self-esteem.

Juvenile crime is at levels once unimaginable. Weapons detectors are routine at school entrances. When you teach students that they're animals, they start acting like animals. "Survival of the fittest" pretty well describes high school society. Evolution is a poor role model: If every organism is trying to win a survival contest, why value another's life? It's a prescription for selfishness as well as brutality.

Anyone familiar with the Old Testament or Torah knows that when ancient Israel honored God, it prospered. But in wealth, it forgot God, disobeyed his commandments, and then declined. (Sound anything like America?) When the Israelites reached the point of ruin, they humbled themselves and cried out to God, and were again blessed. This cycle repeated itself throughout Israel's history.

Of course, that's baloney to the atheist, since the Bible is only "myths and legends." So let's call on some modern wise men. Will Durant, author of *The Story of Civilization*, was one of the preeminent historians of our time. Shortly before his death, he said:

> By offering evolution in place of God as a cause of history, Darwin removed the theological basis of the moral code of Christendom. And the moral code that has no fear of God is very shaky. That's the condition we are in.[8]

Nobel Prize winner Alexander Solzhenitsyn said of his native Russia:

> But if I were asked today to formulate as concisely as possible the main cause of the ruinous revolution that swallowed up some 60 million of our people, I could not put it more accurately than to repeat: "Men have forgotten God; that's why all this has happened."[9]

I think most people would agree that America's big transformation came during the 1960s. In 1963, we were watching Ward Cleaver tell Beaver the importance of honesty. Four years later, we were snapping our fingers as Grace Slick sang about tripping on drugs. Quick transition.

Many people still regard the fifties and early sixties very nostalgically. In fact, they already longed for it by 1973, as evidenced by the success of *American Graffiti*, and the TV show it helped inspire, appropriately named *Happy Days*. Nineteen sixty-three was bygone America's twilight year. Vietnam was on the threshold. So were the Beatles, who would dominate the 1964 charts. Rock 'n' roll was about to turn from happy to heavy, as Don McLean later lamented in "American Pie." Even SAT scores reflected the change. Beginning in 1963, they went down for eighteen consecutive years.[10] What happened that year?

Many thought it was the war, or perhaps John F. Kennedy's death, that triggered the new era of violence, lost innocence and crumbling national character. Personally, I suspect those were symptoms more than root causes.

I don't want to oversimplify matters, but if there was a pivot, I believe it was June 17, 1963. That was the day the U.S. Supreme Court in effect renounced God, banishing him from America's schools. In 1962, the Court had already declared unconstitutional a prayer read in the New York public schools. The "offensive" material was as follows:

> Almighty God, we acknowledge our dependence upon Thee, and we beg Thy blessings upon us, our parents, our teachers, and our country.

Meanwhile, atheist Madalyn Murray O'Hair pursued a lawsuit on behalf of her young son William, protesting prayer and Bible reading at his Baltimore school. The Supreme Court decided that resentment by a few atheists outweighed the free speech of millions of believers. Beginning in 1963, all children in public schools were forbidden an expressed relationship with God. Did drugs, sex and hard rock move in to fill a void? Were two decades of falling SATs a coincidence?

There was another factor. Prior to the 1960s, evolution had not been forcefully pushed in American schools. A number of Southern states still had statutes forbidding its teaching. This restrained the evolutionary content of biology textbooks, which had to market in those states.[11]

Then, in 1959, the 100th anniversary of *The Origin of Species*, the University of Chicago held a Darwinian Centennial Celebration. Speaking at the event, Julian Huxley declared:

> In the evolutionary pattern of thought there is no longer either need or room for the supernatural. The earth was not created, it evolved. So did all the animals and plants that inhabit it, including our human selves, mind and soul as well as brain and body. So did religion.[12]

That year, the National Science Foundation, a U.S. government agency, granted $7 million to the Biological Sciences Curriculum Study (BSCS), which began producing high school biology textbooks with a strong evolutionary slant. Given taxpayer funding, market considerations were no longer a worry. In the 1960s, public schools started using BSCS textbooks. In the meantime, surviving Southern anti-evolution laws were repealed or struck down by the Supreme Court.

Students of the sixties thus faced a two-edged sword. On one hand, they were taught evolution, which effectively repudiated God and the Biblical version of creation. On the other hand, Supreme Court rulings prohibited teachers from discussing God, reading from the Bible, or praying. It was legal to *deny* God's existence, but *illegal* to affirm it.

One reason Americans accepted the imbalance was the stench *Inherit the Wind* had given Christianity. Who wanted to be associated with its grotesque caricatures of fundamentalists? Religion was now unfashionable and passé; a Biblical Christian about as welcome as a dead rat at a party.

Given these factors, the sixties experience was not a mystery, but a logical, predictable outcome.

The Supreme Court's duty is not to make laws—that rests with our elected representatives. Its responsibility is simply to determine if laws accord with the Constitution. In recent years, however, many have charged Court members with judicial activism—overturning legislation because they disagree with it *ideologically*. Such actions are rather easy to justify by misconstruing the Constitution. The modern Court's long assault on religion is a good example.

That assault continued with its 1980 decision banning the Ten Commandments from classrooms, and culminated with its 1992 edict prohibiting ministers from saying prayers at graduation ceremonies. Oh, great, kids are shooting each other, they're getting gorked on crack, and what's the Supreme Court worried about? That somebody might teach them "Thou shalt not kill" or—horrors—say the word "God" at graduation. Now, there are *real* threats to society. We can all sleep well knowing the Court is out there protecting us with astute priorities.

The Constitution's First Amendment states that "Congress shall make no law respecting an establishment of religion, or *prohibiting the free exercise thereof*." The Court has used the Establishment Clause to excuse its anti-God decrees, while simply ignoring the Free Exercise Clause.

Sure, no one wants a dictatorship by a particular religious group; that was why the Constitution's authors wrote the Establishment Clause. Denominations had historically localized in various of the colonies—Puritans in Massachusetts, Baptists in Rhode Island, Episcopalians in Virginia, Catholics in Maryland, Quakers in Pennsylvania. No group wished to be ruled by another.

But they all had a common denominator: Christianity. Since religion was not banned in *any* American school in 1789, you don't need a jurisprudence degree to figure out that the Constitution was never intended to prohibit it. John Jay, first Chief Justice of the U.S. Supreme Court, was also president of the American Bible Society. Judeo-Christian doctrines are the foundation of Western civilization.

Kick God out, and we learn, as the ancient Israelites did, that everything collapses. Since God didn't exist, we decided the rules of marriage were needless. Family life then began disintegrating. In the sixties, we thought everything would be golden after "liberation" from God. "If it feels good, do it," we said. A few thousand ODs and a few million VD cases later, that advice rings a little hollow. Evolution says we're getting better, but in so many ways we're worse.

We had another saying back then: "Make love, not war." But ironically, when sexual passions go out of control, violent ones seem to also. Spiraling violence came hand in hand with the sexual revolution of the sixties. No self-control means no self-control. I don't want to count the times I've read about another child beaten to death by mom's live-in boyfriend.

We used to rage about those square, inhibited, fascist Victorians who would weigh us down with morality. But you know what? The Puritans of old New England went for eighty years without a single murder or rape.[13] Hm—maybe morality *does* have a couple of plusses.

And of course, the academic performance of American students has generally become disgraceful. College professors face freshmen who can't spell, locate Mexico on a map, or identify who Adolf Hitler was. The major teacher's union, the National Education Association, keeps saying that a few more billion dollars into the system will cure all. But that's already been tried. America spends more per pupil than any other nation on Earth, and underfunded foreign students still whip us on standardized exams.

Learning has declined in proportion to *discipline*. It isn't surprising that teenagers whose school plies them with condoms shun studying. If led to indulge their sexual impulses, they're also going to indulge their lazy impulses, no matter how many IBM compatibles they're given. They'll kick those monitors right in.

In 1785, Thomas Jefferson wrote to his nephew Peter Carr, then about to embark on a course of studies: "[Read] everything in the original and not in translations. First read Goldsmith's history of Greece. . . . Then take up ancient history in the detail, reading the following books in the following order: Herodotus, Thucydides, Xenophontis hellenica, Xenophontis Anabasis, Quintus Curtius, Justin. This shall form the first stage of your historical reading. . . ."[14] Jefferson continued at length, making many comparable recommendations.

Nor was this an isolated case of a youth prodigy. But who today boasts of nephews that read Greek and Roman classics in the original language? Why has the American intellect declined so far? Could it be that our forefathers' religious faith, and the self-discipline attained therefrom, had some relation to their capacity for learning?

Ironically, William Murray, the very boy whose case resulted in the Supreme Court's ban on school prayer, is today a devout Christian. He has summed the ordeals that brought him to faith in a book entitled *My Life Without God*. Nearly two decades after the Court's decision, he wrote in a letter published in the *Baltimore Sun*:

> I would like to apologize to the people of the City of Baltimore for whatever part I played in the removal of Bible reading and praying from the public schools of that city. I now realize the value of this great tradition and the importance it has played in the past in keeping America a moral and lawful country. I can now see the damage this removal has caused to our nation in the form of loss of faith and moral decline.[15]

Maybe the rest of us should take a cue.

Plate 74. Clarence Darrow's defense of killers Leopold and Loeb (seated behind him, Loeb's face partly obstructed) contradicted his argument in the Scopes trial, where he minimized the impact of teaching on moral behavior.

Plate 75. Julian Huxley, apostle of evolution, declared that "there is no longer either need or room for the supernatural."

Plate 76. William J. Murray now heads the William J. Murray Evangelistic Association.

CHAPTER 20

Good Company

Inherit the Wind falsely portrayed John Scopes as persecuted and imprisoned, leaving the impression that creationists oppose academic freedom and evolutionists favor it. Events of recent years belie this image.

A 1981 poll commissioned by NBC and the Associated Press revealed that 86 percent of Americans favor allowing creationism to be taught in schools.[1] Nevertheless, the courts gave evolutionists the upper hand. Although Tennessee, Louisiana and Arkansas passed "balanced treatment" laws mandating that evidences for creation receive as much consideration as evolution, by the early 1980s high courts had struck them all down.

In 1980, Lloyd Dale, who had been a biology teacher in South Dakota for seventeen years, was dismissed for teaching both creation *and* evolution as possible explanations for the origin of life.[2]

In 1983, in Tennessee—Scopes country—a mother named Vickie Frost was arrested and jailed after trying to remove her daughter from a class that taught evolution and other materials that violated her religious beliefs. She later won a civil rights suit for false arrest.[3]

In 1996, Aaron Mason, a teacher in Washington State, was suspended for two days without pay for showing his eighth-graders a video on creation and arranging a speaker to present scientific facts favoring creation.[4]

That same year, a Lakewood, Ohio, high school senior wrote an editorial for his school paper, praising physics teacher Mark Wisniewski, who had encouraged students to consider how their own worldviews affected their outlook on the creation-evolution debate. The student said Wisniewski "wanted us to make up our own minds rather than spoonfeed us like other educators." No students or parents complained, but when word leaked, out-of-town calls flooded the school district, the ACLU threatened litigation, and Wisniewski was ordered to stop.[5]

Also in 1996, Danny Phillips, a Denver fifteen-year old, wrote a paper criticizing a *Nova* program he had been required to watch. *Nova* had presented evolution as unquestionable, but Phillips pointed out the lack of evidence. When administrators, impressed by Danny's paper, considered withdrawing the program, the furor from Darwinists was so great that the *Denver Post*'s editorial page editor wrote: "These defenders of intellectual freedom behaved, in fact, just like a bunch of conservative Christians. Theirs was a different kind of fundamentalism, but no less dogmatic and no less intolerant."[6]

Phillip Johnson notes of this last case: "The uproar so upset science educators that they brought out a really big gun to squelch the high-school student. Bruce Alberts, president of the National Academy of Sciences, personally responded to Danny in an editorial published in the *Denver Post*."[7] How the Academy had changed. Twenty years earlier it had issued an *Affirmation of Freedom of Inquiry and Expression* that read, in part:

> I hereby affirm my dedication to the following principles:
>
> . . . That the search for knowledge and understanding of the physical universe and of the living things that inhabit it should be conducted under conditions of intellectual freedom, without religious, political or ideological restriction.
>
> . . . That all discoveries and ideas should be disseminated and may be challenged without such restriction.
>
> . . . That freedom of inquiry and dissemination of ideas require that those so engaged be free to search where their inquiry leads, free to travel and free to publish their findings without political censorship and without fear of retribution in consequence of unpopularity of their conclusions. Those who challenge existing theory must be protected from retaliatory reactions.[8]

The pressure is not limited to public schools. Creationists sometimes have difficulty earning degrees because university programs demand an evolutionary viewpoint. When Robert Gentry was working toward a physics doctorate at Georgia Tech, his research into radiohalos suggested a young Earth. Gentry was told to pursue "a more conventional thesis topic."[9] Later, after testifying on creationism's behalf at the Arkansas "balanced treatment" court case, his research contract with Oak Ridge National Laboratory was discontinued.[10]

Tenured professor Dean Kenyon taught a course on evolution at San Francisco State University. But when he coauthored *Of Pandas and People*, a book advocating "intelligent design," he was censured and his course taken away.[11]

Dr. Halton Arp, considered one of the world's finest astronomers, was president of the Astronomical Society of the Pacific in the early 1980s.

However, after disputing the interpretation that "red shift" substantiates the Big Bang, Arp was denied access to telescopes and finally moved to Europe to continue his work.[12]

The discrimination extends to publishing, where major science journals generally constitute an "establishment" that brooks no dissent with evolution. Hannes Alfvén, who shared the 1970 Nobel Prize for physics, found it nearly impossible to get cosmology articles published after he repudiated the Big Bang.[13]

Michael Behe recounts the unfortunate experience of Forrest Mims:

> In 1990 *Scientific American* asked a science writer named Forrest Mims to write several columns for the "Amateur Scientist" feature of their magazine. . . . The understanding was that if the editors and readers liked the columns, Mims would be hired as a permanent writer. The trial columns all went very well, but when Mims came to New York for a final interview he was asked if he believed in evolution. Mims replied, well, no, he believed in the biblical account of creation. The magazine refused to hire him.[14]

Evolutionists commonly charge that creationists don't publish articles in leading science journals. This is untrue; many have published in the standard periodicals in their field of expertise. However, these scientists commonly find they must mask any hint of creation to obtain editorial approval.

In 1985, Roger Lewin claimed in *Science* that no evidence for creation could be found in technical literature.[15] When Dr. Robert Gentry responded, documenting that he had published numerous articles on polonium halos (which validate creation)—including several in *Science* itself—the magazine simply refused to print his letter. Dr. D. Russell Humphreys of Sandia National Laboratories then wrote, noting that if the journal wouldn't even publish creationists' *letters,* it was hardly surprising that they didn't exert more effort submitting articles. *Science*'s response? They wouldn't print Humphreys' letter either.[16] Thus the denial that creation scientists publish became a sort of self-fulfilling prophecy.

Even some non-creationists are fed up with the discrimination. The accomplished physicist J. Willits Lane wrote in a letter appearing in *Physics Today*:

> After reading a spate of virulently anti-creationist articles and letters in your publication, I decided that something less virulent and more thoughtful should be said. . . .
>
> I have myself sat in class after class in the sciences and humanities in which any idea remotely religious was belittled, attacked, and shouted down in the most unscientific and emotionally cruel way. I have seen

> young students raised according to fundamentalist doctrine treated like loathsome alley cats, emotionally torn apart, and I never thought that this sort of treatment was any better than the treatment that religious prelates, who held authority, gave Galileo. Why scream about the inhumanity of nuclear war if you are also willing to force people of fundamentalist faiths to attend public schools in which their most cherished beliefs will be systematically held up to ridicule and the young children with it? These people are mostly too poor for private schools to be an alternative. The state tries to prevent them from teaching their children at home rather than sending them to school. What choices do they have? Would you call it freedom? Do you call it fair?
>
> Is it really a terrible thing for a textbook to mention that, aside from the Darwin theory of evolution, there have existed other ideas, many of them religious in nature? Would that not open the mind of students rather than close them to scientific possibilities?[17]

When Pieter G. W. J. van Oordt retired as professor of biology at the University of Utrecht in the Netherlands, he remarked during his farewell address:

> To call Christians and Jews stupid and—with a modern abusive word—"fundamentalistic," and even to laugh at them because of their faith in the truth of the Biblical account of creation, is a disgrace to biology.[18]

John Scopes, as portrayed in *Inherit the Wind*, has a modern counterpart in the Christian believer. Today, it is evolution that is taught as dogma and penalizes nonconformists. Students are told Darwinism is an indisputable fact—even though Darwin himself only called it a theory. Many people currently believe in evolution simply because they have never been exposed to any other perspective.

Richard D. Alexander, professor of zoology at the University of Michigan and an evolutionist, has pointed out:

> No teacher should be dismayed at efforts to present creation as an alternative to evolution in biology courses; indeed, at this moment creation is the only alternative to evolution. Not only is this worth mentioning, but a comparison of the two alternatives can be an excellent exercise in logic and reason. Our primary goal as educators should be to teach students to think, and such a comparison, particularly because it concerns an issue in which many have special interests or are even emotionally involved, may accomplish that purpose better than most others.[19]

But as we have seen, attempts at balanced treatment have been shut down—creation science is "religion," so it "violates separation of church and state." The logic is faulty, however. Creation science does not rely on the Bible, but science alone. Some claim it has a hidden agenda of "bringing religion in through the back door," but evolution could just as easily be charged with bringing *atheism* through the back door.

Contrary to popular assumptions, the phrase "separation of church and state" does not appear anywhere in the Constitution. The words are borrowed from an 1802 letter Thomas Jefferson wrote to a group of Connecticut ministers. Ironically, he was assuring the pastors that the government would in no way interfere with *them*—that was the phrase's original context: protecting church from state, not vice versa.

The evolutionary establishment fears creation science because evolution itself crumbles when challenged by evidence. In the 1970s and '80s, hundreds of public debates were arranged between evolutionary scientists and creation scientists. The latter scored resounding victories, with the result that, today, few evolutionists will debate. Isaac Asimov, Stephen Jay Gould, and the late Carl Sagan, while highly critical of creationism, all declined to debate.[20]

A tactic frequently used to invalidate creation scientists is denying they are scientists at all: why, no real scholar would even contemplate a designed universe. It is even hinted that such thoughts reveal insanity. For example, at the 1959 Darwinian Centennial Celebration in Chicago, Garrett Hardin stated that anyone who failed to honor Darwin "inevitably attracts the psychiatric eye to himself."[21] *Inherit the Wind* ingrained this image by portraying both William Jennings Bryan and a fictitious preacher as having nervous breakdowns while babbling Bible verses.

Well! I guess scientific minds don't believe in God or creation. Let's check it out.

What shall we say of Isaac Newton (1642–1727), who discovered the law of gravity, formulated the three laws of motion, developed calculus, constructed the first reflecting telescope, and whom many consider the greatest scientist who ever lived?

Newton wrote an estimated 1,400,000 words on religion—more than on physics or astronomy.[22] He wrote papers refuting atheism and defending the Bible; he believed in the Flood, a literal six-day creation, and the Ussher chronology (which dated Earth as a few thousand years old).[23] Here are a few quotes from him:

> I have a fundamental belief in the Bible as the Word of God, written by men who were inspired. I study the Bible daily.[24]

> All my discoveries have been made in answer to prayer.[25]

> We account the Scriptures of God to be the most sublime philosophy. I find more sure marks of authenticity in the Bible than in any profane history whatsoever.[26]

William Jennings Bryan, for making similar affirmations, was depicted as irrational and anti-scientific in *Inherit the Wind.*

How about astronomer Johannes Kepler (1571–1630)? Reasoning that the universe must be orderly if designed by God, he discovered the laws of planetary motion and conclusively demonstrated that the sun is the solar system's center. He explained that he was merely "thinking God's thoughts after Him"[27] and said:

> I had the intention of becoming a theologian . . . but now I see how God is, by my endeavors, also glorified in astronomy, for "the heavens declare the glory of God."[28]

And:

> Since we astronomers are priests of the highest God in regard to the book of nature, it befits us to be thoughtful, not of the glory of our minds, but rather, above all else, of the glory of God.[29]

What of Robert Boyle (1627–1691), regarded as the father of modern chemistry, and whose name is wedded to the fundamental law of gas pressures? He determined that gases consist of particles, made early discoveries concerning vacuums, and even invented the first match.

Boyle also read the Bible daily, was governor of a missionary organization, wrote *The Christian Virtuoso* to show that studying nature is a religious duty, and in his will established the "Boyle lectures" for the proving of Christianity.

Then there was Francis Bacon (1561–1626), credited with developing the scientific method. He said:

> There are two books laid before us to study, to prevent our falling into error; first, the volume of the Scriptures, which reveal the will of God; then the volume of the Creatures, which express His power.[30]

How about Blaise Pascal (1623–1662), the brilliant French mathematician who developed the science of hydrostatics and helped formulate the laws of probability? From 1658 until his death, he worked on a defense of Christianity. He said:

> Except by Jesus Christ we know not what our life is, what our death is, what God is, what we are ourselves. Thus, without Scripture, which has only Jesus Christ for its object, we know nothing, and we see only obscurity and confusion in the nature of God, and in nature herself.[31]

Hm! Sounds like something a narrow-minded fundamentalist bigot would say.

Biologist John Ray (1627–1705) was first to suggest classifying organisms by species, and was considered the leading authority on zoology and botany in his day. He also wrote a number of theological books, including *The Wisdom of God Manifested in the Works of the Creation.*

We previously mentioned Carolus Linnaeus (1707–1778), who laid the foundations of modern taxonomy, still known as the Linnaean system. He too was a believer. Isaac Asimov acknowledged that "Linnaeus himself fought the whole idea of evolution stubbornly."[32] The *Dictionary of Scientific Biography* says of him:

> His view of nature was deeply religious; central to all his work was God's omnipotence. . . . "I saw," he wrote in the later editions of *Systema natura*, "the infinite, all-knowing and all-powerful God. . . . I followed his footsteps over nature's fields and saw everywhere an eternal wisdom and power, an inscrutable perfection."[33]

Astronomer Sir William Herschel (1738–1822) discovered Uranus and built the greatest reflecting telescopes of his day. He said: "The undevout astronomer must be mad."[34] His son, John Frederick Herschel, who discovered more than 500 stars and nebulae, declared:

> All human discoveries seem to be made only for the purpose of confirming more and more strongly the truths that come from on high and are contained in the sacred writings.[35]

John Flamsteed (1646–1719), who made the first great map of the stars, was founder of the famous Greenwich Observatory, first Astronomer Royal of England—and a clergyman.

The Roman Catholic Church condemned Galileo (1564–1642) for upholding that the planets rotated around the sun (the church's Earth-centered cosmology had been adopted from Ptolemy's teachings, not the Bible). Nevertheless, Galileo was no atheist. "From the Divine Word," he said, "the Sacred Scripture and Nature did both alike proceed."[36]

Besides being a great statesman, Benjamin Franklin (1706–1790) invented the lightning rod, rocking chair, Franklin stove, and bifocal glasses, and coined the terms "positive and negative charges," "battery," and "conductor." He organized the first U.S. postal service and first fire department. Some count Franklin as an unbeliever, but although he poked fun at dour ministers, and entertained some doubts about the divinity of Jesus Christ, his belief in God was uncompromising. He stated:

> Here is my creed. I believe in one God, the Creator of the universe. That he governs it by His Providence. That he ought to be worshipped. That the most acceptable Service we render to him is in doing good to his other Children. That the soul of Man is immortal, and will be treated with Justice in another Life respecting its conduct in this.[37]

He declared before the Constitutional Convention:

> I have lived, Sir, a long time, and the longer I live, the more convincing proofs I see of this truth—that God governs in the affairs of men. And if a sparrow cannot fall to the ground without his notice, is it probable that an empire can rise without his aid?. . .
>
> I therefore beg leave to move—that henceforth prayers imploring the assistance of Heaven, and its blessing on our deliberations, be held in this assembly every morning before we proceed to business. . . .[38]

Some evolutionists like to call creationists "flat earthers." This is ironic since Christopher Columbus, famed for showing the world round, wrote:

> I prayed to the most merciful Lord about my heart's great desire, and He gave me the spirit and the intelligence for the task: seafaring, astronomy, geometry, arithmetic, skill in drafting spherical maps and placing correctly the cities, rivers, mountains and ports. I also studied cosmology, history, chronology, and philosophy.
>
> It was the Lord who put into my mind (I could feel His hand upon me) the fact that it would be possible to sail from here to the Indies. All who heard of my project rejected it with laughter, ridiculing me. There is no question that the inspiration was from the Holy Spirit, because he comforted me with rays of marvelous illumination from the Holy Scriptures. . . .[39]

Shall we add to the list of believers Cotton Mather (1663–1728), the clergyman/Harvard president who introduced a smallpox inoculation; Jean Deluc (1727–1817), the Swiss naturalist who coined the word "geology"; or James Parkinson (1755–1824), the first physician to recognize the dangers of a perforated appendix, and to describe the disease named for him? We could also mention John Dalton (1766–1844), who revolutionized chemistry by developing the atomic theory; Benjamin Barton (1766–1815), who wrote the first U.S. textbook on botany; and chemist-physiologist William Prout (1785–1850), who was first to identify basic foodstuffs as fats, proteins and carbohydrates. And does the famous painting *The Last Supper* not convey the faith of Leonardo da Vinci (1452–1519), considered by many the founder of modern science?

I hear atheists grumbling, "OK, OK, maybe a few of those old dudes had some smarts, but they were only religious because that was the prevailing view in their day. They lived *before* Darwin. If they had read *The Origin of Species,* they would have seen things totally different."

Yeah, right! Newton, awed by the bigger-brained Darwin, would have chucked his Bible out the window and gone atheist. Oh, here he comes down the street now, with his Walkman and his "S**t happens" t-shirt on. Slap me five, Isaac baby!

But hold on. Most scientists in Darwin's time weren't thrilled with his theory, either. Contrary to the popular impression, it was scientists—not theologians—who primarily opposed evolution in the nineteenth century. The Catholic Church, still smarting from its wrongful condemnation of Galileo, wanted no risk of another embarrassment. Although the church maintained an index of forbidden books, *The Origin of Species* and *The Descent of Man* were never placed on it. When Darwin died, the Anglican Church even insisted he be given a hero's funeral and state burial at Westminster Abbey.[40]

On the other hand, 717 scientists, including 86 members of the Royal Society (Britain's most prestigious scientific organization), signed a manifesto entitled "The Declaration of Students of the Natural and Physical Sciences." Issued in London in 1864, it affirmed their confidence in the Bible's scientific integrity.[41]

At the close of Chapter Ten, we quoted several scientists whom evolution failed to convince. There were many others whose faith it could not shake. Louis Pasteur (1822–95) probably saved more lives than any other scientist. He established the germ theory of disease and the process of sterilization; he isolated pathogens and developed vaccines to combat them—including rabies, diphtheria, and anthrax. Of course, he introduced milk pasteurization, which is named for him.

Pasteur was also a humble Christian. He did not patent his discoveries, but gave them to society freely. Though tragedy marked his life—three of his children died young—faith sustained him. "Science," he said "brings man nearer to God."[42] And he observed: "The more I study nature, the more I stand amazed at the work of the Creator."[43] In a series of experiments, Pasteur disproved the false notion, then pushed by evolutionists, that bacteria "spontaneously generate."

Lord Kelvin (1824–1907) was, according to the *Encyclopaedia Britannica*, "foremost among the small group of British scientists who helped to lay the foundations of modern physics."[44] He established a scale of absolute temperatures, with degrees "kelvin" named for him; supervised laying of the first Atlantic cable, for which he was knighted; held 21 honorary doctorates, published more than 600 scientific papers, and patented 70 inventions.

As Chairman of England's Christian Evidence Society, Lord Kelvin said:

> I have long felt that there was a general impression in the non-scientific world that the scientific world believes Science has discovered ways of explaining all the facts of nature without adopting any definite belief in a Creator. I have never doubted that that impression was utterly groundless.[45]

Kelvin opposed Darwinism and published a paper refuting uniformitarian geology. He said: "Overwhelmingly strong proofs of intelligent and benevolent design lie around us . . . the atheistic idea is so non-sensical that I cannot put it into words."[46] And:

> Mathematics and dynamics fail us when we contemplate the earth, fitted for life but lifeless, and try to imagine the commencement of life upon it. This certainly did not take place by any action of chemistry, or electricity, or crystalline grouping of molecules under the influence of force, or by any possible kind of fortuitous concourse of atoms. We must pause, face to face with the mystery and miracle of creation of living creatures.[47]

James Clerk Maxwell (1831–1879) made numerous contributions to mathematics and physics. He developed the kinetic theory of gases, and an electromagnetic field theory that paved the way for radio and television as well as Einstein's theory of relativity. Einstein called Maxwell's achievement "the most profound and most fruitful that physics has experienced since the time of Newton."[48] Henry Morris wrote of him:

> His Christian beliefs were essentially "fundamentalist" in nature. He was strongly opposed to evolution and was able to develop a rigorous mathematical refutation of the famous "nebular hypothesis" of the French atheist LaPlace. He also wrote an incisive refutation of the evolutionary philosophies of Herbert Spencer, the great advocate of Darwinism. A prayer found in his handwriting after his death quoted the Genesis account of man's creation in God's image and the command to subdue the earth as the motivation for his own scientific studies, while also acknowledging his personal faith in Jesus Christ as Lord and Savior.[49]

John Ambrose Fleming (1849–1945), professor of electrical engineering at the University of London for over forty years, pioneered modern electronics. He devised the first electron tube, and invented the thermionic valve—a vital part of radios and TVs until finally replaced by the transistor. He was knighted for his scientific achievements and received many other honors.

An active Christian, Fleming also wrote a major book opposing evolution, and led what was then called the Evolution Protest Movement in Britain. London's *The Times* reported in February 1935:

> Sir Ambrose Fleming presided at a crowded meeting held at Essex Hall, Essex Street, Strand, last evening to launch a public protest against "the teaching of organic evolution as a scientific truth."...
>
> Sir Ambrose Fleming said that of late years the Darwinian anthropology had been forced on public attention by numerous books or highly illustrated periodicals in such fashion as to create a belief that it was a certainly settled scientific truth, and any objections to it were treated as the result of ignorance or bigotry. The fact that many eminent naturalists did not agree that Darwin's theory of species production had been sufficiently established as a truth was generally repressed. If there had been no creation, there was no need to assume any Creator, and the chief basis for all religion was taken away and morality reduced to mere human expediency.[50]

Alexander MacAlister (1844–1919), professor of anatomy at Cambridge, authored many leading textbooks on zoology and physiology. He wrote:

> I think the widespread impression of the agnosticism of scientific men is largely due to the attitude taken up by a few of the great popularizers of science, like Tyndall and Huxley. It has been my experience that the disbelief in the revelation that God has given in the life and work, death and resurrection of our Savior is more prevalent among what I may call the camp followers of science than amongst those to whom scientific work is the business of their lives.[51]

Joseph Lister (1827–1912) saved countless lives by developing antiseptic surgery through the use of disinfectants. ("Listerine" is named after him.) He invented dissolving stitches and the wiring of broken bones. He was knighted, made president of the Royal Society and president of the British Association for the Advancement of Science.

Lister was the son of devout Quakers. Biographer Richard B. Fisher noted of his youth: "He became deeply religious. Indeed, he displayed that most un-Quakerlike attribute, religious enthusiasm."[52] Although later more restrained in expressing his faith, Lister would declare: "I have no hesitation in saying that in my opinion there is no antagonism between the Religion of Jesus Christ and any fact scientifically established."[53]

Samuel F. B. Morse (1791–1872) invented the telegraph and Morse Code, built the first camera in America, and founded the National Academy of Design. He was also an accomplished portrait artist and sculptor.

A dedicated Christian, Morse established one of America's first Sunday schools and supported missionaries. He said:

> The only gleam of hope, and I can not underrate it, is from confidence in God. When I look upward it calms my apprehensions for the future, and

> I seem to hear a voice saying: "If I clothe the lilies of the field, shall I not also clothe you?" Here is my strong confidence, and I will wait patiently for the direction of Providence.[54]

And he said:

> The nearer I approach to the end of my pilgrimage, the clearer is the evidence of the divine origin of the Bible, the grandeur and sublimity of God's remedy for fallen man are more appreciated, and the future is illumined with hope and joy.[55]

The first message he sent by telegraph was: "What hath God wrought."

Physicist and chemist Michael Faraday (1791–1867) developed the sciences of electricity and electromagnetism, invented the generator and transformer, pioneered the liquefaction of gases, and discovered benzene. In *New Scientist,* Jim Baggott wrote of Faraday:

> He was a devout member of the Sandemanian Church, a fundamentalist Christian order that demanded total faith and total commitment. Sandemanians organized their daily lives through their literal interpretation of the Bible. . . . Faraday found no conflict between his religious beliefs and his activities as a scientist and philosopher. He viewed his discoveries of nature's laws as part of the continual process of "reading the book of nature," no different in principle from the process of reading the Bible to discover God's laws. A strong sense of the unity of God and nature pervaded Faraday's life and work.[56]

In his biography of Faraday, L. Pearce Williams wrote:

> His true humility lay in a profound consciousness of his debt to his Creator. . . . In a very real sense, Faraday's science was firmly rooted in his faith. . . . What higher goal could a man seek than the knowledge of God's creation? In this way, he could participate, however infinitesimally, in the Divinity.[57]

Faraday never accepted Darwin's theory. An elder in his church for over twenty years, he said:

> The Bible, and it alone, with nothing added to it nor taken away from it by man, is the sole and sufficient guide for each individual, at all times and in all circumstances. . . . Faith in the divinity and work of Christ is the gift of God, the evidence of this faith is obedience to the commandment of Christ.[58]

Matthew Maury (1806–1873), known as the "pathfinder of the seas," wrote the first textbook of modern oceanography. He became the leading authority on ocean currents and prevailing winds, and the charts he developed significantly reduced ocean travel time. Annapolis's main academic hall is named for him.

The Bible inspired Maury's work. It refers to "the paths of the sea," and Maury set out to find them. He said: "The Bible is true and science is true, and therefore each, if truly read, but proves the truth of the other."[59]

James Simpson (1811–1870), professor of obstetric medicine at Edinburgh University, discovered chloroform's use as an anesthetic, and helped innovate the field of gynecology. He said:

> But again I looked and saw Jesus, my substitute, scourged in my stead and dying on the cross for me. I looked and cried and was forgiven. And it seems to be my duty to tell you of that Saviour, to see if you will not also look and live.[60]

Through experiments, James Joule (1818–1889) proved the law of energy conservation, and determined the mathematic relationship between an electric current's energy and the heat it gives off. The *joule*, a unit of energy measurement, is named for him. He said: "It is evident that an acquaintance with natural laws means no less than an acquaintance with the mind of God therein expressed."[61]

Though born a slave, George Washington Carver (1864–1943) became one of the world's greatest agricultural scientists. Working at the Tuskegee Institute, an Alabama school for Afro-Americans, he developed over 300 products from the peanut and 118 from the sweet potato. He showed both black and white farmers how to better utilize land, and revitalized the South's economy. He did much to improve race relations, and was also an accomplished artist.

Like Pasteur, Carver patented none of his discoveries, but gave them away. He turned down an offer from Thomas Edison to leave Tuskegee and work at sixty times his pay. In 1940, he donated his life savings to the Institute. A devout Christian, Carver taught his students from the Bible, in a class that met on Sundays from 1907 until his death.[62] He said:

> The secret of my success? It is simple. It is found in the Bible, "In all thy ways acknowledge Him and He shall direct thy paths."[63]

The great French biologist Henri Fabre (1823–1915), who pioneered modern entomology, received his country's ribbon of the Legion of Honor. He said of God:

> Without Him I understand nothing; without Him all is darkness. . . . Every period has its manias. I regard Atheism as a mania. It is the malady of the age. You could take my skin from me more easily than my faith in God.[64]

So to figure out that science harmonizes with the Bible, you don't have to be a rocket scientist—but you might ask one. Wernher von Braun (1912–1977) was director of NASA's space flight center; he oversaw the team of scientists that sent the first American into space, and masterminded the moon landing.

An active Christian, von Braun prayed for the safety of those on the manned missions he planned.[65] He observed: "There are those who argue that the universe evolved out of a random process, but what random process could produce the brain of man or the system of the human eye?"[66] He would not have agreed with the whip hand given evolution in today's classrooms: "To be forced to believe only one conclusion—that everything in the universe happened by chance—would violate the very objectivity of science itself."[67]

The *first* men to fly, Wilbur and Orville Wright, were also Christians. Biographer Charles Ludwig related that their father, a bishop in the United Brethren Church, "never tired of relating the positive effect that the Bible had on his children."[68]

Did I forget mathematician Charles Babbage (1792–1871), who invented the first actuarial tables, first speedometer, first skeleton keys, first ophthalmoscope, and first train "cowcatcher"? He also designed calculating machines recognized as forerunners of the computer. A committed believer, he wrote a treatise on the compatibility of science and revealed religion. It included a mathematical analysis of Biblical miracles.

Physicist David Brewster (1781–1868) began the science of optical mineralogy, invented the kaleidoscope, and was founder and president of the British Association for the Advancement of Science. He said: "Knowledge, indeed, is at once the handmaid and the companion of true religion. They mutually adorn and support each other." One of his students recalled: "He thanked God that the way of salvation was clear and simple; no labored argument, no hard attainment was required. To believe in the Lord Jesus Christ was to live; he trusted Him and enjoyed His peace." On his tombstone was written: "The Lord is my Light."[69]

Shall we add Joseph Henry (1797–1878), who invented the electromagnetic motor and galvanometer, was first secretary and director of the Smithsonian Institution, president of the American Association for the Advancement of Science—and always prayed for divine guidance during any experiment?[70] Or Sir William Huggins (1824–1910), the astronomer who first demonstrated that stars consist mostly of hydrogen, and who was a professing Christian? Or Nobel Prize winner Lord Rayleigh (1842–1919), co-discoverer of argon, helium, and the other "noble" gases, who wrote:

"The works of the Lord are great, sought out of all them that have pleasure therein."[71] Nobel Prize-winning physicist Robert A. Millikan (1868–1953), though critical of fundamentalism, said he would still choose it over atheism: "The God of science is the Spirit of rational order, and of orderly development. Atheism as I understand it is the denial of the existence of this spirit. Nothing could be more antagonistic to the whole spirit of science."[72] Need I mention that Gregor Mendel (1822–1884), who formulated the laws of heredity, was abbot of a monastery? Many other eminent scientists could join our list.

Some may consider the men named in this chapter intellectual midgets compared to today's agnostic and atheistic evolutionists, such as Stephen Jay Gould and Isaac Asimov. But the latter have had no monopoly on modern times; many legitimate scientists are still Bible believers. The Missouri-based Creation Research Society has over 600 voting members with post-graduate science degrees, each fully committed to creation instead of evolution. All the faculty of California's Institute for Creation Research hold advanced scientific degrees. In 1997, Dr. Raymond Damadian, inventor of the medical diagnostic device known as the MRI, joined the Institute's Technical Advisory Board.

Among all the scientists who ever lived, atheists compose a small minority. Clearly, no one is an "enemy of science" for believing the Bible. What sad irony that Isaac Newton, applying for a job today at *Scientific American,* would be turned down cold if the editors applied the same standard to him as to Forrest Mims.

PLATE 77. "When I look at the solar system, I see the earth at the right distance from the sun to receive the proper amounts of heat and light. This did not happen by chance."—Isaac Newton

PLATE 78. "I find it as difficult to understand a scientist who does not acknowledge the presence of a superior rationality behind the existence of the universe as it is to comprehend a theologian who would deny the advances of science."—Wernher von Braun

PLATE 79. Johannes Kepler

PLATE 80. Samuel F. B. Morse

PLATE 81. Louis Pasteur

PLATE 82. Gregor Mendel

PLATE 83. "Why, then, should we who believe in Christ be so surprised at what God can do with a willing man in a laboratory? Some things must be baffling to the critic who has never been born again."
—George Washington Carver

CHAPTER 21

You and the Man Upstairs

If evolutionists honor this book with any critical reviews, one of their charges may be that I don't hold an advanced degree. That's perfectly true. All I have is a Bachelor of Science from Boston University. "And so," the review may go, "why should anyone accept this man's opinions over the authority of distinguished scientists with doctorates?"

That is one reason I wrote the last chapter—to demonstrate the multitude of scientists agreeing with this book's central position. Yes, evolution may be the "prevailing view" in America right now, but on the other hand, Nazism was once the prevailing view in Germany. Trendy is not necessarily the same as true.

The evolution-creation debate embraces many topics: biology, molecular biology, chemistry, biochemistry, paleontology, paleoanthropology, geology, genetics, physics, astronomy, mathematics, radiometric dating, taxonomy, zoology, etc. No person holds degrees in all these fields—if that was required to speak on evolution, everyone would be silent. A Darwinist, by reason of a doctorate in one or even two of these areas, does not qualify as an expert in the rest. A specialist in one subject is usually a layman in another. You, the reader, by reviewing the facts, are competent to draw conclusions about evolution, whether or not you have a Ph.D. in paleobotany.

Furthermore, denying visible evidence based on the "authority" of evolutionists contradicts the very thesis of *Inherit the Wind*—that men should be allowed to think for themselves, unimpeded by dogma. To support its case, this book *has* cited experts in each field addressed. One authority, however, I have not yet appealed to.

In Job 38:4, God has a question for evolutionists: "Where were you when I laid the earth's foundation?" Who saw the Big Bang explode, or Earth formed from gas and dust? Who ever saw life develop spontaneously, or mutations generate new organs, or one species transform into another? A science should be based on supporting observations, but evolution is virtually barren of them. In any other field, if facts didn't fit a theory, it would be discarded. Evolution survives because it is, at its root, an attempt to deny God. Its opposition to faith is ironic, since, with so little corroborating evidence, it must be accepted on faith itself.

Still, the atheist rails that reason and faith are incompatible—eternal enemies, with reason embracing science, and faith embracing superstitious mumbo-jumbo. That misrepresents the Bible, which frequently exalts wisdom and knowledge. (I didn't realize this as an atheist, because I "knew all about" the Bible without studying it.) It declares, for example:

> Wisdom is supreme; therefore get wisdom.
> Though it cost all you have, get understanding.[1]

In the Bible, God says, "Come now, let us reason together."[2] Faith isn't incompatible with reason—evolution is. The error has been in confusing Darwinism with "science." Unfortunately, many Christians have done just that, and attempted to reconcile their beliefs with it. They have watered Christianity down and become "theistic evolutionists," assuming that God created the world billions of years ago and let life happen by chance and evolution.

Taking this view, however, makes faith meaningless. If everything evolved, then morality and even our religious beliefs evolved. Some Christians adopted theistic evolution hoping that atheists and Darwinists would meet them halfway and accept God. But they deceived themselves, for atheists correctly see evolution as antithetical to theology. As G. Richard Bozarth noted in *American Atheist*:

> Christianity has fought, still fights, and will fight science to the desperate end over evolution, because evolution destroys utterly the very reason Jesus' earthly life was supposedly made necessary. Destroy Adam and Eve and the original sin, and in the rubble you will find the sorry remains of the son of God. If Jesus was not the redeemer who died for our sins, and this is what evolution means, then Christianity is nothing.[3]

We're through discussing the Darwin side of the conflict. If evolution isn't true, and the world and humanity are designed, then we're facing God.

Two hundred years ago, watching the King of England ride by, some people would sigh, "Ah, if only I could live in a palace like him, and enjoy all his comforts, I would be happy."

Today, we average folks are much better off than that King of England. He couldn't telephone anyone, flick on electric lights, go to the movies, fly in airplanes, play a stereo, surf the Net, turn on air conditioning, or taste citrus fruits flown in during winter. His doctors knew less about health than the informed lay person of today.

Yep. Seems we outclass the guy who once was everyone's envy. But people still aren't very happy, are they? Important as material things are, they don't bring lasting, meaningful contentment. Many who remember the 1950s will attest that we were generally happier then, even though we were technologically less advanced. The title of TV's *Happy Days* required no explanation.

I'm speaking broadly, but in those days, you walked city streets at night with little risk of being mugged. You sent your kid to the store without worrying that her face would end up on milk cartons. In many regions, people didn't bother locking their doors. If you moved into a new home, neighbors turned out with pastries to greet you. School entrances had no weapons detectors. Illicit drugs: what were they? Things weren't perfect, but they were sure better. Personally, I'd ditch my VCR any day to resume that social climate.

We wouldn't have to heave our VCRs, though, because technology had nothing to do with it. Our nation's happiness was no mystery; it was linked to faith. I personally believe the tide turned in 1963 when the Supreme Court repudiated God. Before then, as long as we honored him, he blessed us.

As a rule, our ancestors were happier because their ultimate aim was not wealth, health, or success—which are obtainable, but never guaranteed to last. They were happier because their goal was heaven: a pursuit that disaster, disease, and death itself cannot take away. They pinned their hopes to the spiritual, not the material.

Heck, there were people in the 1800s who schemed to become millionaires, too. But even the few who made it are dead now. The real question is, where are they today?

As an old hippie, do I speak only for myself in saying, sometimes it's not that people *don't* believe in God, but *won't*? If someone had shown me a book like this when I was eighteen, I would have chucked it and said, "Ah, bunch of bull#*%#!" God meant morality—ugh, the big "M." We were the sixties generation. We rejected anything old, as if age somehow made an idea less true. We had lots of slogans, and they haven't died: "You can't legislate morality." "If it feels good, do it." "There are no absolutes." "Do your own thing." "There is no objective truth—truth is relative." We didn't want anyone preaching about "sin" and telling us how to run our lives, especially the sex part. But are these logical reasons for denying God's existence?

Sure, the Ten Commandments include lots of "don'ts." So do the instructions that come with your microwave. You don't want to break it—or your body. And how many bodies have been damaged by alcohol, cigarettes, drugs and VD? Sumpin' wrong with a little common sense? That's what the Bible calls for. God's not against sex—he invented it. Pleasures are not evil, but their abuse is. "If it feels good, do it" stunk as advice. Any pleasure, even video games, becomes addictive and unhealthy if overdone.

Unfortunately, some people spend their lives trying to prove wrong is right and vice versa. I should know—I was one of them. We justified this by saying "right and wrong" weren't fundamental, but dogmas taught by our particular society—if we happened to live in another society, it might teach us a totally different set of morals. I remember me and the other eighteen-year old geniuses endlessly yapping about such matters in philosophy class at Colby College.

But these ideas break down before hard analysis. Values have actually been rather similar from culture to culture. Courage and honesty are universally liked, for example; selfishness is universally disliked. Yes, morals are taught, but that doesn't make them untrue, any more than teaching a multiplication table lessens its validity.

To say there are no absolutes is a contradiction in logic, for it is an absolute statement itself. If there is no right and wrong, can we be sure we are right about that? "There is no objective truth"—but is that an objective truth? "Truth is relative"—but is that only relatively true?

Even a person who denies the existence of good and evil becomes angry if you break a promise to them. People *do* have an inherent sense of right and wrong. In 1996 in Scotland, a gunman walked into a school and killed sixteen kindergartners. Does anyone want to tell their parents that wasn't wrong? Or argue that Mother Teresa's service to the poor was by comparison deplorable?

Saying there is no right and wrong may sound sophisticated in philosophy class, but just doesn't cut it in the real world. Standards exist. Love *is* better than hate. And if there are betters, there should be bests. The highest best, the perfect ideal, would be represented by God himself.

In nearly every culture, people have worshipped something greater than themselves. This universality infers that life means more than molecules. Even atheists validate it in a way—why do many of them hate God so fervently if he doesn't even exist? They usually despise only the God of the Bible, while indifferent to other religions' gods.

It is objected that since we don't *see* God, he isn't there. But God is not part of his creation. We earthlings live in just a small corner of it, over a comparatively short time. The Bible says that, for God, "a thousand years are like a day." If we entered the Empire State Building, and inspected one room, should we conclude that, because we didn't see the architect, the building never had one, and must have arisen by chance?

Not everything real is visible and quantifiable—that's one limitation of physical sciences. Who can measure love? You might be thinking of purple, but you'd have a hard time proving it. Test tube experiments can't define beauty or morality. As astrophysicist Arthur Stanley Eddington said, "You can no more analyze these imponderables by scientific methods than you can extract the square root of a sonnet."[4] Intangibles, like courage and kindness, often seem to matter *most* in life. The Bible says God judges us by our hearts. And if we can accept the subjective, many people affirm that they *have* seen God, experiencing his presence in their lives.

Is it religious dogma to acknowledge God's reality? If so, excluding God as a possibility seems no less dogmatic. Who comprehends the universe well enough to state with certainty, "There is no God"? Who can prove the claim?

The problem of evil

Perhaps the most common objection to God's existence: If he is perfect and loving, why is there so much evil? There are essentially two kinds: (1) moral evil: murder, cruelty and every injustice; (2) pain and suffering, both physical and mental. If God created a world full of these, isn't he evil himself? This problem has several answers. We won't use "God works in mysterious ways" as one of them.

Some pain is recognizably useful. As we reach toward a flame, pain warns us not to touch.

And growth necessitates some unpleasantness. Our children may find homework uncomfortable; that doesn't mean we should do it for them. In the same way, God may permit us to endure some difficulties so we can learn and mature. Many people have attested that they drew wisdom, character or strength from a time of suffering.

Failure, for example, is painful. But we become better people if we learn from it. Mistakes teach us what doesn't work. It's pretty difficult to become wise without knowing any struggle or defeat.

Some suffering is self-inflicted: the spendthrift who finds himself broke, the drunk who gets d.t.'s, the adulterer facing divorce.

But it doesn't always work that way, does it? Plenty of pain comes for no apparent reason. Good people may suffer tragedy, while scoundrels catch breaks. Hard work usually produces rewards—but some folks see labors end in ruin, while a ne'er-do-well wins the lottery.

The Bible says of creation: "God saw all that he had made, and it was very good." Wasn't God wrong on that point? Seems the world's got plenty that *ain't* good. This observation helped drive Darwin from Christianity and into evolution. Charles, however, missed on his theology.

The main answer to the problem of evil lies in freedom. People want freedom to choose everything—their spouses, careers, friends, beliefs, political leaders, houses, cars, clothes, food.

God created man with free will. Now some people disagree, arguing that we are just products of heredity, or react to the environment on a stimulus-response basis. But if free will doesn't exist, why do we encourage others to change their minds? Sure, sometimes emotions rule us, and we go out of control—but we also reflect, weigh, and choose.

God created free will—not evil. If I hit you with a rock, it's not the rock that's malicious, but the decision to use it that way. However, power to choose is not evil in itself.

A mother tells her child to clean up his room. Later she finds it's a mess. This wasn't the mother's intent. Quite the opposite. But because the child had free will, this outcome was possible. Likewise, God did not decree that a Hitler or Stalin commit their genocides. These acts were defiance against God.

Why, then, did God create free will, if evils have resulted? Perhaps because without it, our spirits are dead. We cherish freedom; historically, there is nothing people have been more willing to die for. We don't *wish* to be robots. God doesn't want us that way either—he desires that we obey him of our free will. But true freedom cannot exist unless choosing evil is an option.

Moral wickedness aside, what about pain and suffering? Obviously, no one chooses that. The Bible doesn't say the world is perfect *today*—it was perfect *when God created it*. Adam and Eve lived in Eden, a paradise without sickness, suffering or death. Things collapsed when Adam and Eve sinned. The Bible says God had warned them that they would "surely die" if they chose disobedience. Death then entered the world, both physically and spiritually, and creation itself became cursed.

This set in motion one of nature's most fundamental laws: the Second Law of Thermodynamics, which Isaac Asimov summarized as follows:

> "The universe is constantly getting more disorderly!" Viewed that way, we can see the Second Law all about us. We have to work hard to straighten out a room, but left to itself it becomes a mess again very

> quickly and very easily. Even if we never enter it, it becomes dusty and musty. How difficult to maintain houses, and machinery, and our own bodies in perfect working order; how easy to let them deteriorate. In fact, all we have to do is nothing, and everything deteriorates, collapses, breaks down, wears out, all by itself and that is what the Second Law is all about.[5]

Anything declines if unattended. A glass of boiling water cools off; a twirled ball stops spinning. The sun is burning out; even the Earth's rate of rotation is measurably slowing. Oil and natural gas supplies will eventually be exhausted. Our clothes, cars and TVs wear out. Bodies, though they temporarily grow under a genetic blueprint, also wear out and die. According to the Second Law, the entire universe is winding down. (This is, incidentally, another axiom of science that mitigates against evolution, which views everything as *building up*.) As the Bible puts it: "In the beginning you laid the foundations of the earth, and the heavens are the work of your hands. They will perish, but you remain; they will all wear out like a garment."[6]

Why do good people experience pain, suffering, death, or persecution? First, let's face this: Who is truly good? No one leads a perfect life. We all have an inherent tendency to sin and rebel; since Adam and Eve, it's been transmitted from generation to generation.

Moreover, the *comparatively* good people still suffer because we live in a run-amuck world that rebelled against God and became cursed. And isn't the rebellion still going on? We're flouting the Bible and its commandments; we've told God to buzz off so we can do it "our way." We have our wish—God is standing aside.

The Bible says that it rains on the just and unjust alike.[7] Leading good and productive lives results in many rewards—but doesn't guarantee we won't encounter evil people, bodily pain or death, because all creation is under the curse. So even with free will, we are only partial, not complete, masters of our fate. In *Forrest Gump*, Forrest summed it pretty well when he told his friend Jenny: "I don't know if we each have a destiny, or if we're all just floating around accidental-like on a breeze, but I think maybe it's both—maybe both happening at the same time."

The world is not *all* suffering, is it? It holds much beauty and pleasure. We still have reminders of what Eden was like. In our souls, we wish to restore that time. For do we not seek perfect happiness? People rarely yearn for something with no corresponding reality. We wouldn't feel hunger if there wasn't food, or lust without sex. So our desire for utopian bliss seems to validate Eden, for if we were really animals raised from the muck, and perfection had never existed, where would we acquire such a longing? As Peter Kreeft and Ronald K. Tacelli put it:

> We behave as if we remember Eden and can't recapture it, like kings and queens dressed in rags who are wandering the world in search of their thrones. If we had never reigned, why would we seek a throne? If we had always been beggars, why would we be discontent? People born beggars in a society of beggars accept themselves as they are. The fact that we gloriously and irrationally disobey the first and greatest commandment of our modern prophets (the pop psychologists)—that we do *not* accept ourselves as we are—strongly points to the conclusion that we must at least unconsciously desire, and thus somehow remember, a better state.[8]

The road home

According to the New Testament and gospel of Jesus Christ, the door is not shut forever. "Gospel" means "good news." There is a way back to Eden—each individual has an opportunity for eternal life and happiness. It requires a decision to come to God.

For those unfamiliar with the Biblical background, we will encapsulate it. As we have stated, God created the world in a perfect state, and man with free will. Man chose to disobey and rebel, separating himself from God, and originating evil, suffering and death.

But God kept reaching out. The Old Testament explains that he chose the Jews as his people, to set an example for the world. Through Moses, he gave them the Ten Commandments and other rules to live by. He required that they sacrifice the best of their livestock and grains as atonement for sins. He gave them Israel to live in, and greatly blessed them. However, he also warned the Jews that if they disobeyed him and forsook his laws, they would be conquered and scattered throughout the Earth. That is exactly what happened.

But God told the Jews he would never completely forsake them. While they survived, and Israel was reestablished in 1948, the peoples they competed with in the ancient Middle East—the Philistines, Babylonians, Assyrians, Hittites, Phoenicians, etc.—all disappeared as cultures. Clearly, the Israelites possessed something superior.

Meanwhile, no one was perfectly obedient to God, keeping the commandments and being truly good. People were too infected with evil; they continued to sin and suffer. Mankind needed a savior.

In *Beauty and the Beast*, a prince was placed under an evil spell (in the Disney version, as punishment for an arrogant and cruel act). He became a hideous beast, and the curse could only be broken by a maiden who loved him despite his ugliness. When this finally happened, he was saved, and restored to his former kingdom. That story is not unlike the Biblical account of sin and salvation.

People could not overcome sin by themselves—only God could do that. Therefore, in his chosen land, Israel, God clothed himself in flesh and became a man—his own son, Jesus Christ. Israel's prophets had foretold his coming for centuries. Under self-imposed humanity, he knew hunger, pain, and other sufferings of mankind. But he was perfect in love, and without sin—and thus could defeat the ultimate consequences of sin: sickness and death. Everywhere, he poured out compassion, healing the ill and crippled, even raising the dead, confirming his divinity. Previously, God had commanded the Jews to make sacrifices to atone for sin. Now he would make the greatest sacrifice—his own son, as restitution for *mankind's* sins. Jesus Christ was tortured and crucified—nailed to a cross. While dying, he bore the punishment for man's evil.

With that, the curse was broken and death overcome. Christ demonstrated this by his Resurrection: three days after the crucifixion, he rose again, appearing to hundreds of people. That is the New Testament's central message. It did not mean suffering and death simply ended—for the world was still sinful. It meant reconciliation with God was now freely available: that by turning to him and accepting Christ's atoning sacrifice, our sins would be forgiven, and we could live with God in heaven, again knowing Eden. That is the gospel's "good news." "The wages of sin are death," says the Bible, "but the gift of God is eternal life through Jesus Christ our Lord."[9]

Around this point, the atheist interjects: "But there's no evidence that Jesus Christ ever even existed!" Actually, he is one of history's best-documented figures, recorded not only in the New Testament's various eyewitness accounts, but noted by Roman and Jewish historians. If we dismiss his reality because of "insufficient documentation," we must, on that basis, throw out thousands of other historical figures accepted on far less evidence. Disliking what someone says is a poor reason for denying their existence. How can a man, whose birth has been the basis for figuring the calendar date throughout the world, never have even lived?

"Aw," the atheist says, "Jesus was just a lie the apostles made up." But the apostles and other early Christians were persecuted: imprisoned, tortured, crucified, beheaded, and fed to lions. Who would endure that for a lie?

Some critics, while admitting Jesus existed, insist that fables eventually developed about him, accounting for the reported miracles. But the gospels are written matter-of-factly. While mythic tales seek to glorify some ancient figure, the gospels reveal Jesus' human side, the apostles' faults, and minor details of everyday life—features normally missing from aggrandizing legends. Although the Romans destroyed Jerusalem in A.D. 70—less than forty years after the crucifixion—the gospel writers knew the city intimately, demonstrating that they were contemporaries of the events described.

The gospels' validity was uncontroversial in the church, which accepted them as factual from the beginning. It is true that stories tend to change when transmitted verbally. That's why the gospels were committed to writing. Altering those accounts would have been considered a flagrant sin. The Bible's last chapter solemnly admonishes:

> I warn everyone who hears the words of the prophecy of this book: If anyone adds anything to them, God will add to him the plagues described in this book. And if anyone takes words away from this book of prophecy, God will take away from him his share in the tree of life and in the holy city, which are described in this book.

In ancient Israel, religious scribes were extremely careful when copying Scriptures. If even one mistake was found, the entire manuscript would be thrown out. This does not mean minor transcription errors never occurred, but it made them very rare. The discovery of the Dead Sea Scrolls in 1947 confirmed this. Prior to then, the oldest known copy of the Old Testament dated to around A.D. 1000. The Dead Sea Scrolls included portions of every book except Esther. Transcribed from about 200 B.C. to A.D. 70, they proved the faithfulness of later copies: variations were minimal after a thousand years.

We have New Testament fragments dating from the second century A.D. By contrast, the oldest surviving copy of Caesar's *Gallic Wars* was made one thousand years after first being written, yet no one doubts its authenticity. If we're going to reject the Bible because the documents aren't old enough, we better discard other ancient literature as well.

Furthermore, archaeology has consistently corroborated both the Old and New Testaments. Numerous kings, cultures and places they describe, once discounted as mythical by Bible critics, have been validated in excavation after excavation. Dr. Nelson Glueck, president of Hebrew Union College and considered by many the leading Palestinian archaeologist of the twentieth century, said that "it may be stated categorically that no archaeological discovery has ever controverted a Biblical reference. Scores of archaeological findings have been made which confirm in clear outline or in exact detail historical statements in the Bible."[10]

Some say, "Christ may have existed, but he just deceived everyone, using magician's tricks that looked like miracles. Ancient people, being superstitious airheads, swallowed it." But how could illusions heal people blind or crippled since birth? Nor were the ancient Jews gullible. Most of their leaders regarded Jesus skeptically. Furthermore, deceivers usually seek power and wealth. Jesus Christ lived a completely unselfish, nonmaterialistic life; he willingly suffered torture and death—hardly your average con man's ambitions.

Some say Christ deluded *himself*—a lunatic who thought he was divine. True, there are residents of mental institutions who say they are God. But in meeting such people, you quickly sense their irrationality. Christ, on the other hand, was a profound moral instructor; his wise teachings have been a creed for millions. In his own day, many religious scholars tried to rebut him in debate—all failed. These are not earmarks of deranged self-deception.

Some people won't accept Christianity because they don't buy miracles. "Miracles violate natural law, I've never seen one, they aren't happening today, and why should I accept the Bible's authority on them?" That something isn't a norm doesn't mean it can't happen. You wouldn't tear up a winning lottery ticket because the odds against were ten million to one. The Bible doesn't say miracles are everyday occurrences—God used them primarily to underscore certain events, such as the Jewish exodus from Egypt, and Christ's coming.

Few of us have seen an electron or the planet Pluto; none of us saw George Washington cross the Potomac. We accept the word of authorities that such things exist or happened. The Bible is an authority, too, and a reliable one. The creation of the universe itself—whether we advocate the Big Bang or creation—violated natural law. God can intervene in the world he made, and break the rules when he pleases. No one has disproven the miracles.

Is there evidence for Christianity? Dr. Simon Greenleaf, Dane Professor of Law at Harvard University in the mid 1800s, wrote *A Treatise on the Law of Evidence*, widely regarded as the greatest authority on evidence in legal literature. Chief Justice Fuller of the U.S. Supreme Court called Greenleaf "the highest authority cited in our courts,"[11] and the *London Law Journal* stated: "Upon the existing law of evidence (by Greenleaf) more light has shone from the New World than from all the lawyers who adorn the courts of Europe."[12]

Greenleaf was also Jewish and, initially, an agnostic who considered the Resurrection a hoax. Irritated by students who quoted Scripture, Dr. Greenleaf accepted a challenge to use standard rules of evidence to determine if the Resurrection was an historical event. He penned an exhaustive study, *An Examination of the Testimony of the Four Evangelists by the Rules of Evidence Administered in Courts of Justice*, in which he concluded: "It was therefore impossible that they [the apostles] could have persisted in affirming the truths they have narrated, had not Jesus actually risen from the dead, and had they not known this fact as certainly as they knew any other fact."[13] He later wrote:

> Of the Divine character of the Bible, I think no man who deals honestly with his own mind and heart can entertain a reasonable doubt.[14]

Lew Wallace, diplomat and Civil War general, was also an unbeliever. "I had no conviction about God and Christ,"[15] he later recalled. But after investigating the Bible to see if it was true, he became a Christian and penned *Ben Hur: A Tale of the Christ*, the basis for one of Hollywood's greatest movies.

In the sixties, we big bad rebels prided ourselves on fighting "the system." Later I realized we'd been working for it all along. The drugs and other sins that enslave people—*that's* the system. We done been had, baby. Jesus Christ and the early Christians, well, maybe they didn't wear leather jackets, but they certainly were the original rebels, fighting the system even to their deaths.

If you're an unbeliever, I invite you to do what Greenleaf, Wallace, and millions of others, including myself, have done: reevaluate your position. God wants you on his team. How about playing for the good guys? You just might like it.

Maybe you feel too morally jaded to embrace God, but if so, you weren't always that way—perhaps you can at least remember childhood moments when you weren't. Why not return to your father's house? Even if you never knew a loving father on Earth, there's a heavenly father who does love you, and is waiting.

What does becoming a Christian mean? First, it *doesn't* mean becoming a narrow-minded bigot who passes judgement on others. A few Christians may act that way, but there's often a large gap between a faith and some who practice it. Jesus condemned religious hypocrisy and warned: "Do not judge, or you too will be judged." Occasionally, there have also been financially scandalized preachers, and priests who sexually abused choir boys. But such behavior does not exemplify Christianity—it violates it. Human beings are fallible, and may disappoint you, but that is not God.

Some claim Christianity is mean-spirited. On the contrary, its teachings' essence is love, forgiveness, truthfulness and humility. Christ commanded us to love even our enemies, and "Do to others what you would have them do to you." Meanwhile, evolution postulates survival of the fittest—a prescription for selfishness.

Atheists have also accused Christians of being escapists, shunning life's reality for "pie in the sky." But are not atheists, by denying God, seeking to escape accountability to moral standards?

Christianity is unique; among the founders of world religions, only Jesus performed miracles and claimed to be God. Some therefore dislike Christianity because upholding one religion over others seems undemocratic. It violates our American sense of equality and fairness. But while everyone is assuredly equal in their right to hold views, those views are not necessarily equal in truth.

It's been said that all religions reveal God, merely differing in viewpoint, just as blindfolded people touching an elephant would give various impressions, because some felt its trunk, some its tusks, and some its tail. But while other religions may possess significant truth, they are often irreconcilable with Christianity. For example, Hinduism maintains that we lead multiple lives, being reincarnated after death. The Bible declares that we are appointed to live but once, and then face God. These beliefs are mutually exclusive.

Christianity says that God made us; the New Age says that we *are* God. But if man is the highest being possible, the world's in trouble. Claiming we're God is like Hamlet asserting he's the playwright, or the picture that it's the painter. In fact, the original sin and fall of man embodied this illusion. In Genesis 3:5, Satan tempted Eve to eat from the tree of the knowledge of good and evil, promising that "your eyes will be opened, and *you will be like God*." This lie remains the heart of New Age philosophy. Cults riddle our world today, and frankly, it was evolution that opened the door to many of them. After the Bible's long-accepted wisdom was "proven" false, people, being spiritual creatures, began seeking religious truth elsewhere. I was among them.

Many equate Christianity with simply going to church, hearing boring sermons, and dressing up on Easter. Yes, one can be nominally Christian and go through these motions without experiencing God's richness. One reason religion seems lifeless and dull in America: evolution even convinced many churches, eroding beliefs. Some mainstream denominations discarded essential doctrines—the Bible's accuracy, the story of creation, Christ's divinity, the miracles, and the Resurrection. Faith grew increasingly meaningless until some churches became little more than social clubs.

Christianity is not about *religion*; it's about being "born again": experiencing personal regeneration by establishing a relationship with God. It's about you—your soul.

Now some people protest that they have no soul and consist only of molecules. But there is a soul—it's you, the one with an identity who's thinking. Look at a dying person. When he passes away, the molecules are still there, but isn't the corpse missing something profound? One of physics' most basic laws says energy can never be destroyed. At death, then, where does the life go that energized the body? If it cannot be destroyed, it must go somewhere.

Blaise Pascal, the eminent French mathematician, formulated what is known as "Pascal's wager": If you die as a Christian, and it turns out you were wrong—there is no God or afterlife—well, you didn't lose anything by believing. In fact, you were probably happier while alive. But if you were right and there is a God, you inherit heaven and everlasting life.

On the other hand, if you die an atheist, and you were right—there is no God or afterlife—well, you haven't lost anything, and haven't won anything either. But what if you're wrong? You will have forfeited heaven and entered hell. Pascal's point: Christianity offers everything to gain and nothing to lose, atheism nothing to gain and everything to lose.

Did I say "hell"? Yes, the Bible lays that out as clearly as heaven. How could God justly permit hell to exist? Actually, justice seems to demand it. Should Mother Teresa and Adolf Hitler wind up in the same place? If everyone enters heaven automatically, what difference does the kind of life we lead make? Like other consequences of sin, hell is something that we, not God, create. Overeat, you get a bellyache; drive off a cliff, you die. Deny God, hell becomes your destiny. Why? Because that's what hell is—separation from God's love and light.

Cardiologist Maurice Rawlings has written an interesting book, *To Hell and Back* (Thomas Nelson Publishers, 1993). As is well known, people who have near-death experiences—"dying" and then being resuscitated—often tell of seeing heaven, or a great light, or feeling deep inner peace. In his own interviews with resuscitated subjects, however, Dr. Rawlings found that many also encountered the terror of hell during near-death. Why have these experiences been so little publicized? Because the people are reluctant to discuss them, Dr. Rawlings found.

How does one enter heaven? Good deeds won't buy us in, though they should be evident in a born-again life. It's not what we must do, but what Jesus Christ has already done on the cross. Salvation is God's gift, freely available to anyone desiring it. However, more than casual assent is needed. One must sincerely repent of evil he has done, and ask the Lord's forgiveness. God says in the Bible: "You will seek me and find me when you seek me with all your heart."[16] When one does that, and opens his heart to God, God will then give the seeker a new life, changing him from the inside out.

Some people hesitate to become Christians because they fear stereotyped images—that they'll have to become monastery monks or jungle missionaries. But whether you're a housewife or CEO, physicist or cab driver, athlete or artist, God can meet you where you are. Unless you're a practicing criminal, you don't have to surrender your field; God uses people in every walk.

Many also think, "Oh, Christianity is for a goody-two-shoes—I'm not like that." There are no goody-two-shoes. If heaven was for sinless people, it would be empty. Christianity doesn't mean you're above sin—it means *admitting* you're a sinner.

But doesn't faith require giving certain things up? It excludes drunkenness, not alcohol; promiscuity, not sex. One can still make money, but not as one's ultimate goal. Does Seagram's bring happiness? One-night stands end loneliness? Wealth secure fulfillment?

God gives far more than he takes. Some worry Christian life would be boring. Personally, I've had ten times more fun since coming to faith. Living only for material things—nothing's duller than that. Time after time, people have topped the corporate ladder, or reached success in other fields, or bought the house they'd always dreamed of, only to find themselves still miserable and empty inside. Israel's King Solomon was the richest man of his era, yet he declared: "Meaningless! Meaningless! . . . Now all has been heard; here is the conclusion of the matter: Fear God and keep his commandments, for this is the whole duty of man."[17] Only God brings lasting contentment.

Many find that after making God the center of their lives, the things they always wanted and never got come to them anyway. Jesus said, "But seek first his kingdom and his righteousness, and all these things will be given to you as well."[18] And so a person, unable to find a spouse, met one after coming to Christ, because their new sincerity and caring made them far more attractive. And thus a person, having failed in business, succeeded after coming to Christ, because their new self-discipline enhanced their work, and their character impressed employers. Contrary to Leo Durocher's advice, nice guys often do finish first, as the notables listed in Chapter Twenty attest. Biblical principles are a road map for success. George Washington Carver was often heard to say: "The Lord has guided me. He has shown me the way, just as he will show everyone who turns to Him."[19]

Not that coming to faith means perfect behavior. Those who try living as Christians frequently disappoint themselves. Old habits prove hard to conquer. But God forgives failure. As C. S. Lewis pointed out, "Very often what God first helps us toward is not the virtue itself but just this power of always trying again."[20] Born-again Christians have faults, as critics hasten to point out; the real issue is, how much worse would the flaws be *without* Christ?

Becoming a believer doesn't mean things will be easy. Let's face it—life is tough. The world is still under the curse; tragedy afflicts Christian and non-Christian alike. Faith doesn't guarantee wealth or health. It brings happiness, but that doesn't mean constantly *feeling* happy. People exult when their lottery ticket hits, or the home team wins a championship. But such emotions don't last—true happiness is a state, not a feeling. As a Christian, you still experience ups and downs; your *mood* may change (similarly, spouses don't always "feel" in love), but that won't diminish the prize you've won.

Just as a novel's hero and heroine willingly face many challenges so they may live "happily ever after," our eternity matters most. "Weeping may endure for a night," says the Bible, "but joy cometh in the morning."[21] For those in Christ, the curse known since Adam and Eve will vanish. The Bible's final two chapters promise:

> Now the dwelling of God is with men, and he will live with them. . . . He will wipe every tear from their eyes. There will be no more death or mourning or crying or pain, for the old order of things has passed away. . . . No longer will there be any curse. . . . There will be no more night. They will not need the light of a lamp or the light of the sun, for the Lord God will give them light. And they will reign for ever and ever.

You can make a decision for God. You don't need the Supreme Court's approval—God created the universe without their permission (I know it would shock some Court members to learn this). It doesn't matter what Darwinists would think—they can rationalize arguments for evolution, but when they die, they'll face God like everyone who went before them.

It is good to doubt and ask questions—that is seeking truth. But a point comes when a choice should be made. Some say they want 100 percent certainty before committing to God. That is not necessary; a jury normally reaches a verdict without 100 percent certainty. When you leave on a trip, you're not really 100 percent certain you'll arrive, but you go nonetheless.

William Jennings Bryan noted that while evolution would take eons to change people, Christ can transform them in the twinkling of an eye. Perhaps God has a special purpose for your life. I invite you to become, by your own choice, a member of God's family, and receive a new life. Yo, amigo! Let's get back that "Happy Days" feeling of the fifties. Roll the oldies. Come on, play for the good guys again. *Christians* are the counterculture today—we're the underdogs; it's the "Establishment" that's atheist. Being a Christian is strictly hip, way cool.

If you think you might be ready for that change, I encourage you to say the following prayer, not as a recitation, but expressing it from the heart, speaking to God as you would to a person:

Lord, I don't know if you exist or not.
But I do want to know the truth.
If you are real, please hear my prayer.
I acknowledge that I have not led a perfect life.
I have sinned against you and others.
I ask that you forgive me, by the blood Jesus Christ shed on the cross.
I want to try life your way.
Come, Lord, and live in my heart.
Fill me with your love.
Grant me your companionship and leadership.
Give me a clean, new life.
Help me to walk in your ways, to serve you, and to do what is right.
Accept me into your kingdom for all eternity.
Thank you, Lord, for hearing my prayer.

Coming to God is the most important thing you can do. If you've prayed that prayer sincerely, you ought to celebrate.

And there are things to do next. Faith isn't automatic—it needs to be fed.

(1) Get a copy of the world's number one best-seller since the invention of the printing press. The Bible is where to start learning God's will for your life. The New Testament gospels are a good place to begin. You don't have to read a Bible written in old, hard-to-understand language, such as the King James Version. Modern translations are available; the New International Version is currently in wide use.

(2) Pray regularly. Ask God for wisdom, for answers to your needs, and to know his will for you. Be patient. Some people find God changes them overnight; for others, the process is gradual.

(3) You are now part of God's family; enjoy the fellowship of others on the same journey. Find a healthy, well-balanced church. While no church is perfect, you want one that hasn't abandoned Christianity's core doctrines—the authority of Scripture, divinity of Christ, and reality of his Resurrection.

(4) Seek to apply Biblical principles in your life. Jesus said that in receiving God's forgiveness, we must also forgive others. This doesn't mean true forgiving comes easily or instantly, or that we tolerate evil against us; but willingness to forgive is essential to Christian character.

What's most important? When people are on their deathbed, their biggest regret is never, "Gee, I wish I had put in more time at the office." It's usually something like, "I wish I hadn't refused to speak to my son for the last ten years." Follow the golden rule. Treat people as if they were wearing a big sign that says "Make me feel important." Treat them as you would if you knew they were going to die tomorrow.

(5) Avoid the occult—astrology, psychic hotlines, Ouija boards, New Age. Such things may seem fascinating, but any "insight" they provide is not only inconsistent, they can, in time, mess with your head. The Bible forbids occult practices. Comparing them to *Star Wars*' "dark side of the Force"—deceptively attractive, ultimately entrapping—is not totally facetious. Dump 'em.

(6) If you have chosen to go with God, you may have many questions. There are excellent resources that can help you understand the Bible and live a Christian life. Details are in Appendix Two.

On the other hand, perhaps all this means nothing to you. You still find evolution more persuasive than God. That is your free choice. Good luck.

But remember. "The princess kissed the frog, and he turned into a handsome prince." We call that a fairy tale. Evolution says frogs turn into princes, and we call it science.

We say that a pool of nonliving, unthinking chemicals will, given enough time, produce: laughter and love; morality and a sense of justice; sex and human consciousness; tulips, kittens, and ticklish babies; chocolate malts and bagfuls of gummi bears; Mona Lisa, the *1812 Overture*, and the last scene in *Breakfast at Tiffany's*. All these will result because enough grotesque mutations turn out right, preserved by wild beasts clawing each other to death in survival's jungle.

Is that science? Or is it, like the fraud of Piltdown Man, the forgeries of Haeckel's embryos, the misrepresentations of *Inherit the Wind*, and the coercions of the Supreme Court, merely part of a long effort to deny God? Even after all our disobedience, he has not closed his door to us. Two thousand years ago, following a star to a manger, three wise men sought Christ. Wise men still do.

APPENDIX ONE

Creationists Cited in this Book

This list includes only *contemporary* creation scientists and creationist writers whom this book has quoted. Creation scientists of the past, such as Newton and Maxwell, are enumerated in Chapter Twenty.

Austin, Steven, A.
Coffin, Harold G.
Cooper, Bill
Gentry, Robert V.
Gish, Duane T.
Ham, Ken
Humphreys, D. Russell
Johnson, Phillip E.
Kang, C. H.
Lubenow, Marvin
Morris, Henry M.
Nelson, Ethel
Nevins, Stuart E.
Niermann, D. Leland
Oard, Michael J.
Parker, Gary E.
Sarfati, Jonathan
Whitcomb, John C.
Wilder-Smith, A. E.
Woodmorappe, John

APPENDIX TWO

Resources for the New Christian

Note: In recommending resources that are primarily evangelical, I intend no offense to my Catholic friends. Recognizing that today's social crises outweigh the traditional disputes between Catholics and Protestants, I welcome the brotherhood expressed by the "Joint Statement on Justification," issued by theologians from the two bodies, and I hope this book's discussion of creation science will help both, as well as members of Orthodox faiths.

BOOKS

Thousands of edifying books are available to assist your walk with God. In addition to browsing your local Christian bookstore, you can contact:

Christian Book Distributors (CBD)
PO Box 7000
Peabody, MA 01961–7000
(978) 977–5000
www.christianbook.com

Ask CBD to put you on their mailing list. They are the nation's largest distributor of Christian books, and frequently mail out colorful catalogues indexed by topic.

TELEVISION

Many new Christians enjoy *The 700 Club*, featured weekdays on cable's Family Channel. In a variety-show format, it includes inspiring Christian testimonies and Biblical teaching, as well as news and entertainment.

RADIO

Christian radio stations blanket the U.S. and air many outstanding syndicated broadcasts. I recommend *The Bible Answer Man*, featured nationally every weekday. Hank Hanegraaff hosts this informative live call-in show that answers diverse questions about the Bible and Christian faith. Check out *Focus on the Family*, which deals with marriage and child-rearing matters, J. Vernon McGee's *Through the Bible*, which explains the Bible in depth, and D. James Kennedy's *Truths that Transform*, which treats many issues facing today's Christians. If you're having difficulty finding a Christian radio station in your area, a complete national listing, state-by-state with maps, may be obtained by sending $7.00 to Traveler's Companion, PO Box 777, Clinton, MD 20735.

FOR GUIDANCE ON LEADING A CHRISTIAN LIFE

Get in touch with:

The Navigators
PO Box 6000
Colorado Springs, CO 80934
(719) 598–1212
www.gospelcom.net/navs
e-mail: navs@gospelcom.net

CHOOSING A CHURCH

There is no national "clearing house" that recommends specific churches, but for guidelines to help you find a good Bible-based church, try contacting:

The Christian Research Institute
PO Box 7000
Rancho Santa Margarita, CA 92688–7000
(714) 858–6100
www.equip.org

The Institute is also a leading resource for information about cults. Another organization that may be able to assist in finding a church in your area:

The National Association of Evangelicals
PO Box 28
Wheaton, IL 60189
(630) 665–0500
www.nae.net
e-mail: nae@nae.net

FOR COLLEGE STUDENTS

InterVarsity Christian Fellowship has chapters on more than 500 campuses. Their national headquarters is at:

PO Box 7895
6400 Schroeder Road
Madison, WI 53707–7895
(608) 274–IVCF
www.gospelcom.net/iv
e-mail: info@ivcf.org

FOR QUESTIONS OR ADDITIONAL MATERIALS ON CREATION AND EVOLUTION

Contact any of the following:

Institute for Creation Research
PO Box 2667
El Cajon, CA 92021
(619) 448–0900
www.icr.org

Answers in Genesis
PO Box 6330
Florence, KY 41022
(606) 727–2222
www.answersingenesis.org

Creation Research Society
PO Box 8263
St. Joseph, MO 64508–8263
(816) 279–2626
www.creationresearch.org

Scores of books on creation and evolution can be obtained from these organizations. Ask the Institute for Creation Research or Answers in Genesis for one of their colorful catalogues. I want to especially plug Ian Taylor's book *In the Minds of Men*, which uniquely summarizes the history of the creation-evolution debate. Answers in Genesis offers subscriptions to *Creation Ex Nihilo,* a beautiful and very readable magazine dealing with these topics.

FAMILY ISSUES

For resources on marriage and raising children, I recommend, without reservation:

Focus on the Family
PO Box 35500
Colorado Springs, CO 80935–3550
(719) 531–5181
www.family.org

NEWS MAGAZINE

Our spiritual outlook affects every part of our lives, including how we view issues that are, strictly speaking, civil or nonreligious. The mainstream media usually present news with a secular slant, and Christian views are sometimes misrepresented as badly as in *Inherit the Wind.* Mature Christians who seek an alternative, behind-the-headlines look at current events might try the biweekly *The New American.* For subscription information contact:

The New American
PO Box 8040
Appleton, WI 54913
(920) 749–3784

Notes

Impact is a publication of the Institute for Creation Research, El Cajon, Calif.

Chapter Two. Problems Carved in Stone

1. Pierre-Paul Grassé, *Evolution of Living Organisms: Evidence for a New Theory of Living Organisms* (New York: Academic Press, 1977), 4.
2. Carl O. Dunbar, *Historical Geology* (New York: John Wiley and Sons, 1949), 52.
3. Gavin De Beer, "The World of an Evolutionist," *Science* 143 (20 March 1964): 1311.
4. Michael Denton, *Evolution: A Theory in Crisis* (Bethesda, Md.: Adler and Adler, 1986), 189.
5. David M. Raup, "Conflicts Between Darwin and Paleontology," *Field Museum of Natural History Bulletin* 50 (January 1979): 22.
6. Charles Darwin, *The Origin of Species* (1872; reprint, New York: Random House, 1993), 408.
7. Ibid., 406.
8. Richard Owen, "Darwin on the Origin of Species," *Edinburgh Review* 11 (April 1860): 487–532, quoted in David L. Hull, *Darwin and his Critics* (Cambridge, Mass.: Harvard University Press, 1973), 149–50.
9. Darwin, *Origin*, 414.
10. Darwin, *Origin*, 433.
11. Charles Darwin, *Life and Letters*, ed. Francis Darwin, vol. 3 (1888; reprint, New York: Johnson Reprint, 1969), 25.
12. Edmund R. Leach, "Men, Bishops and Apes," *Nature* 293 (3 September 1981): 20.
13. Raup, 22–23, 24–25.
14. Stephen Jay Gould, "Evolution's Erratic Pace," *Natural History* 86 (May 1977): 14.
15. Colin Patterson, letter to Luther D. Sunderland, 10 April 1979, quoted in Luther D. Sunderland, *Darwin's Enigma: Fossils and Other Problems* (San Diego: Master Books, 1988), 89.
16. George Gaylord Simpson, *Tempo and Mode in Evolution* (New York: Columbia University Press, 1944), 107.
17. David B. Kitts, "Paleontology and Evolutionary Theory," *Evolution* 28 (September 1974): 467.
18. Steven M. Stanley, *Macroevolution: Pattern and Process* (San Francisco: W. H. Freeman, 1979), 39.
19. Heribert Nilsson, *Synthetische Artbildung* (Lund, Sweden: Verlag CWK Gleerup, 1953), English summary, 1212.
20. Ibid., 1188.
21. Gareth V. Nelson, "Origin and Diversification of Teleostean Fishes," *Annals of the New York Academy of Sciences* 67 (1969): 22.
22. "Is Man a Subtle Accident?" *Newsweek*, 3 November 1980, 95.
23. Stephen Jay Gould, "A Short Way to Big Ends," *Natural History* 95 (January 1986): 18.

24. Richard Dawkins, *The Blind Watchmaker* (New York: W. W. Norton, 1986), 229.
25. J. R. Norman, *A History of Fishes*, 2nd ed., ed. P. H. Greenwood (New York: Hill and Wang, 1963), 296.
26. Gerald T. Todd, "Evolution of the Lung and the Origin of Bony Fishes: A Causal Relationship?" *American Zoologist* 20 (1980): 757.
27. Robert L. Carroll, "Problems of the Origin of Reptiles," *Biological Reviews of the Cambridge Philosophical Society* 44 (July 1969): 393.
28. Owen, "Darwin on the Origin of Species," quoted in Hull, *Darwin and his Critics*, 211.
29. Denton, 180.
30. Tom Kemp, *Mammal-like Reptiles and the Origin of Mammals* (New York: Academic Press, 1982), 319.
31. Douglas Dewar, "The Case Against Organic Evolution," in *Witnesses Against Evolution*, ed. John Fred Meldau (Denver: Christian Victory Publishing, 1968), 55.
32. W. E. Swinton, "The Origin of Birds," in *Biology and Comparative Physiology of Birds*, ed. A. J. Marshall (New York: Academic Press, 1960), 1.
33. Barbara J. Stahl, *Vertebrate History: Problems in Evolution* (New York: McGraw-Hill, 1974), 350.
34. Peter Farb, *The Insects* (New York: Time, Inc.: 1962), 14, 15.
35. Chester A. Arnold, *An Introduction to Paleobotany* (New York: McGraw-Hill, 1947), 7.
36. Patricia G. Gensel and Henry N. Andrews, "The Evolution of Early Land Plants," *American Scientist* 75 (September/October 1987): 480.
37. E. J. H. Corner, "Evolution," in *Contemporary Botanical Thought*, ed. Anna M. MacLeod and L. S. Cobley (Chicago: Quadrangle Books, 1961), 97.
38. Bruce J. McFadden, *Fossil Horses: Systematics, Paleobiology, and Evolution of the Family Equidae* (Cambridge: Cambridge University Press, 1992), 255, 257–58.
39. Alfred Sherwood Romer, *Vertebrate Paleontology*, 3rd ed. (Chicago: University of Chicago Press, 1966), 260–61.
40. Nilsson, 551–52.
41. Garrett Hardin, *Nature and Man's Fate* (New York: Rinehart, 1959), 260.
42. Boyce Rensberger, "Ideas on Evolution Going Through a Revolution Among Scientists," *Houston Chronicle*, 5 November 1980, sec. 4, p. 15, quoted in Duane T. Gish, *Creation Scientists Answer Their Critics* (El Cajon, Calif.: Institute for Creation Research, 1993), 80.
43. Stahl, 350.
44. Duane T. Gish, "Startling Discoveries Support Creation," *Impact* 171 (September 1987): 4.
45. Francis Hitching, *The Neck of the Giraffe* (New York: Ticknor and Fields, 1982), 34.
46. Jean L. Marx, "The Oldest Fossil Bird: A Rival for Archaeopteryx?" *Science* 199 (20 January 1978): 284.
47. Tim Beardsley, "Fossil Bird Shakes Evolutionary Hypotheses," *Nature* 322 (21 August 1986): 677; Alun [*sic*] Anderson, "Early Bird Threatens Archaeopteryx's Perch," *Science* 253 (5 July 1991): 35.
48. Stephen Jay Gould and Niles Eldredge, "Punctuated Equilibria: The Tempo and Mode of Evolution Reconsidered," *Paleobiology* 3 (Spring 1977): 147.
49. Rensberger, quoted in Gish, *Creation Scientists Answer Their Critics*, 80.
50. Tom Kemp, "A Fresh Look at the Fossil Record," *New Scientist* 108 (5 December 1985): 66.
51. Gould, "Evolution's Erratic Pace,"14.

52. "Fossil Finds," *Geotimes* 33 (May 1988): 27.
53. Duane T. Gish, *Evolution: The Fossils Still Say No!* (El Cajon, Calif.: Institute for Creation Research, 1995), 75.
54. Robert L. Carroll, *Vertebrate Paleontology and Evolution* (New York: W. H. Freeman, 1988), 463.
55. Henry M. Morris and Gary E. Parker, *What Is Creation Science?* (El Cajon, Calif.: Master Books, 1987), 133.

Chapter Three. Marvelous Mutations

1. John Patterson, "Do Scientists and Educators Discriminate Unfairly Against Creationists?" *Journal of the National Center for Science Education* (Fall 1984): 19–20, quoted in Henry M. Morris and John D. Morris, *Society and Creation* (Green Forest, Ark.: Master Books, 1996), 177.
2. Richard Dawkins, "Creation and Natural Selection," *New Scientist* 111 (September 1986): 37.
3. Fritjof Capra, *The Web of Life* (New York: Anchor Books, 1996), 228.
4. James Wynbrandt and Mark D. Ludman, *The Encyclopedia of Genetic Disorders and Birth Defects* (New York: Facts on File, 1991), preface.
5. Gary E. Parker, "Creation, Mutation, and Variation," *Impact* 89 (November 1980): 2.
6. Francis Hitching, *The Neck of the Giraffe* (New York: Ticknor and Fields, 1982), 77.
7. Ernst Chain, *Responsibility and the Scientist in Modern Western Society* (London: Council of Christians and Jews, 1970), 25.
8. Theodosius Dobzhansky, *Genetics and the Origin of Species* (New York: Columbia University Press, 1951), 73.
9. Hitching, 59.
10. Lee Spetner, *Not By Chance!: Shattering the Modern Theory of Evolution* (Brooklyn, N.Y.: Judaica Press, 1997), 131, 138.
11. Michael Pitman, *Adam and Evolution* (London: Rider, 1984), 68.
12. A. E. Wilder-Smith, *The Natural Sciences Know Nothing of Evolution* (Costa Mesa, Calif.: T.W.F.T. Publishers, 1981), 46–47.
13. Michael Behe, *Darwin's Black Box: The Biochemical Challenge to Evolution* (New York: The Free Press, 1996), 41–44.
14. Spetner, 141–42.
15. C. P. Martin, "A Non-geneticist Looks at Evolution," *American Scientist* 41 (January 1953): 101.
16. George Gaylord Simpson, *Tempo and Mode in Evolution* (New York: Columbia University Press, 1944), 54–55.
17. Pierre-Paul Grassé, *Evolution of Living Organisms: Evidence for a New Theory of Living Organisms* (New York: Academic Press, 1977), 103.
18. Spetner, 103.

Chapter Four. Logic Storms Darwin's Gates

1. John H. Ostrom, "Bird Flight: How Did It Begin?" *American Scientist* 67 (January-February 1979): 55.
2. Barbara J. Stahl, *Vertebrate History: Problems in Evolution* (New York: McGraw-Hill, 1974), 349.

3. Stephen Jay Gould, "The Return of Hopeful Monsters," *Natural History* 86 (June/July 1977): 24.
4. Charles Darwin, *The Origin of Species* (1872; reprint, New York: Random House, 1993), 227.
5. Pamphlet no.142, quoted in Francis Hitching, *The Neck of the Giraffe* (New York: Ticknor and Fields, 1982), 98.
6. Michael Pitman, *Adam and Evolution* (London: Rider, 1984), 216.
7. Michael Behe, *Darwin's Black Box: The Biochemical Challenge to Evolution* (New York: The Free Press, 1996), 18–21.
8. Ibid., 194.
9. Ibid., 39.
10. Ibid., 5.
11. Ibid., 136–37.
12. James A. Shapiro, "In the Details…What?" *National Review*, 16 September 1996, 64.
13. Arthur Koestler, *The Ghost in the Machine* (London: Hutchinson, 1967), 129.
14. W. H. Yokel, promotional letter for *Scientific American* (1979), quoted in John Whitcomb, *The Early Earth* (Grand Rapids, Mich.: Baker Books, 1986), 126.
15. Michael Denton, *Evolution: A Theory in Crisis* (Bethesda, Md.: Adler and Adler, 1986), 330.
16. Don DeYoung and Richard Bliss, "Thinking about the Brain," *Impact* 200 (February 1990): 1.
17. Yokel, quoted in Whitcomb, 126.
18. Phillip E. Johnson, *Defeating Darwinism by Opening Minds* (Downers Grove, Ill.: InterVarsity Press, 1997), 81–82.
19. "Drafting the Bombardier Beetle," *Time*, 25 February 1985, 70.
20. Jules Poirier and Kenneth B. Cumming, "Design Features of the Monarch Butterfly Life Cycle," *Impact* 237 (March 1993): 2–3.
21. Lisa J. Shawver, "Trilobite Eyes: An Impressive Feat of Early Evolution," *Science News* 105 (2 February 1974): 72.
22. Riccardo Levi-Setti, *Trilobites*, 2nd ed. (Chicago: University of Chicago Press, 1993), 54, 57.
23. Theodosius Dobzhansky, *Genetics and the Origin of Species* (New York: Columbia University Press, 1951), 4.
24. Austin H. Clark, *The New Evolution: Zoogenesis* (Baltimore: Williams and Wilkins, 1930), 168, 189.
25. W. R. Thompson, introduction to *The Origin of Species*, by Charles Darwin (reprint, New York: Dutton, Everyman's Library, 1956), quoted in Henry M. Morris and John D. Morris, *Science and Creation* (Green Forest, Ark.: Master Books, 1996), 29.
26. Denton, 149.
27. Gavin de Beer, *Homology, An Unsolved Problem* (London: Oxford University Press, 1971), 16.
28. Denton, 249, 250, 278.
29. Hitching, 75.
30. Lee Spetner, *Not By Chance!: Shattering the Modern Theory of Evolution* (Brooklyn, N.Y.: Judaica Press, 1997), 28.
31. Colin Patterson, "Evolution and Creationism," speech at American Museum of Natural History (5 November 1981), cited in Luther D. Sunderland and Gary E. Parker, "Evolution? Prominent Scientist Reconsiders," *Impact* 108 (June 1982): 3.
32. Francisco J. Ayala, "The Mechanisms of Evolution," *Scientific American* 239 (September 1978): 68.

33. Denton, 285.
34. Norman Myers, "Extinction Rates Past and Present," *Bioscience* 39 (January 1989): 39.

Chapter Five. Origin of the Specious

1. Charles Darwin, *The Origin of Species* (1872; reprint, New York: Random House, 1993), 278.
2. Jonathan Weiner, *The Beak of the Finch: A Story of Evolution in Our Time* (New York: Alfred A. Knopf, 1994), 125.
3. Charles Darwin, *The Origin of Species by Means of Natural Selection, or, The Preservation of Favoured Races in the Struggle for Life* (1859; reprint, New York: Avenel Books, 1979), 108. (Note: all other quotes from *Origin* are taken from the Random House reprint of the 1872 edition, cited in footnote 1 above.)
4. *The World's Most Famous Court Trial: Tennessee Evolution Case* (Dayton, Tenn.: Bryan College, 1990), 254.
5. Ibid., 270.
6. Niles Eldredge, "Progress in Evolution?" *New Scientist* 110 (5 June 1986): 55.
7. H. S. Lipson, "Origins of Species," in "Letters," *New Scientist* 90 (14 May 1981): 452.
8. Harold G. Coffin, *Origin by Design* (Hagerstown, Md.: Review and Herald Publishing Association, 1983), 344.
9. Marjorie Grene, "The Faith of Darwinism," *Encounter* (November 1959): 54.
10. Thomas Hunt Morgan, *Evolution and Adaptation* (New York: Macmillan, 1903), 43.
11. Colin Patterson, "Cladistics," interview by Peter Franz, British Broadcasting Corporation television program, 4 March 1982, quoted in Henry M. Morris and John D. Morris, *Science and Creation* (Green Forest, Ark.: Master Books, 1996), 35.
12. Charles Darwin, *Life and Letters*, ed. Francis Darwin, vol. 3 (1888; reprint, New York: Johnson Reprint, 1969), 25.
13. Pierre-Paul Grassé, *Evolution of Living Organisms: Evidence for a New Theory of Living Organisms* (New York: Academic Press, 1977), 130.
14. Francis Hitching, *The Neck of the Giraffe* (New York: Ticknor and Fields, 1982), 55.
15. Norman Macbeth, *Darwin Retried* (Boston: Gambit, 1971), 36.
16. Hitching, 57; Coffin, 407.
17. Darwin, *Origin*, 141.
18. Richard Goldschmidt, "Evolution, As Viewed by One Geneticist," *American Scientist* 40 (January 1952): 94.
19. Coffin, 394.

Chapter Six. Darwin vs. Design

1. Henry M. Morris and Gary E. Parker, *What Is Creation Science?* (El Cajon, Calif.: Master Books, 1987), 33.
2. Cicero, *On the Nature of the Gods*, trans. Horace McGregor (Harmondsworth, England: Penguin Books, 1972), 158–59.
3. Colin Patterson, letter to Luther D. Sunderland, 10 April 1979, quoted in Luther D. Sunderland, *Darwin's Enigma: Fossils and Other Problems* (San Diego: Master Books, 1988), 89.

4. Stephen Jay Gould et al., "The Shape of Evolution: a Comparison of Real and Random Clades," *Paleobiology* 3 (Winter 1977): 34–35.
5. Pierre-Paul Grassé, *Evolution of Living Organisms: Evidence for a New Theory of Living Organisms* (New York: Academic Press, 1977), 8.
6. Ludwig Bertalanffy, "Chance or Law," in *Beyond Reductionism: New Perspectives in the Life Sciences*, ed. Arthur Koestler and J. R. Smythies (Boston: Beacon Press, 1971), 65–67.
7. Charles Darwin, *The Origin of Species* (1872; reprint, New York: Random House, 1993), 579–80.
8. Mark Ridley, *The Problems of Evolution* (New York: Oxford University Press, 1985), 8.
9. Michael Denton, *Evolution: A Theory in Crisis* (Bethesda, Md.: Adler and Adler, 1986), 151–52.
10. Gary E. Parker, "Nature's Challenge to Evolutionary Theory," *Impact* 64 (October 1978): 4.
11. Robert MacArthur, "Population Ecology of Some Warblers of Northeastern Coniferous Forests," *Ecology* 39 (October 1958): 617.
12. Walter Linsenmaier, *Insects of the World* (New York: McGraw-Hill, 1972), 228.
13. Fritz W. Went, "The Ecology of Desert Plants," *Scientific American* 192 (April 1955): 74.
14. Parker, "Nature's Challenge," 1.
15. Gregory Alan Pesely, "The Epistemological Status of Natural Selection," *Laval Theologique et Philosophique* 38 (February 1982): 74.
16. Paul S. Moorhead and Martin M. Kaplan, ed., *Mathematical Challenges to the Neo-Darwinian Interpretation of Evolution,* monograph no. 5 (Philadelphia: Wistar Institute Press, 1967), 14.
17. L. Harrison Matthews, introduction to *The Origin of Species*, by Charles Darwin (reprint, London: J. M. Dent and Sons, 1971), xi.
18. Charles Darwin, *More Letters of Charles Darwin*, ed. Francis Darwin, vol. 1 (London: John Murray, 1903), 450.
19. Charles Darwin, *The Autobiography of Charles Darwin 1809–1882*, ed. Nora Barlow (London: Collins, 1958), 236.
20. Zygmunt Litynski, "Science Around the World," *Science Digest* 50 (January 1961): 61.
21. Arthur Koestler, *Janus: A Summing Up* (New York: Random House, 1978), 185.
22. Norman Macbeth, "A Third Position in the Textbook Controversy," *American Biology Teacher* (November 1976): 495–96.
23. Søren Løvtrup, *Darwinism: The Refutation of a Myth* (New York: Croom Helm, 1987), 422.
24. Sharon Begley, "Science Finds God," *Newsweek,* 20 July 1998, 46, 50.
25. Paul Davies, "The Christian Perspective of a Scientist," *New Scientist* 98 (2 June 1983): 638.
26. Begley, 50.
27. Ibid., 51.
28. Max Planck, "On Religion and Science," appendix A in *The Creation in the Light of Modern Science*, by Aron Barth, trans. L. Oschry (Jerusalem: Jerusalem Post Press, 1968), 144, 148, 149.
29. David Raphael Klein, "*Is* There a Substitute for God?" *Reader's Digest*, March 1970, 55.

Chapter Seven. Vegas Odds on Life

1. Ian T. Taylor, *In the Minds of Men: Darwin and the New World Order* (Minneapolis: TFE Publishing, 1991), 187–89.
2. Charles Darwin, *Life and Letters*, ed. Francis Darwin, vol. 1 (1888; reprint, New York: Johnson Reprint, 1969), 18.
3. John Farley, *The Spontaneous Generation Controversy from Descartes to Oparin* (Baltimore: Johns Hopkins University Press, 1977), 73.
4. "New Evidence on Evolution of Early Atmosphere and Life," *Bulletin of the American Meteorological Society* 63 (November 1982): 1328–29.
5. Ibid., 1329.
6. Harry Clemmey and Nick Badham, "Oxygen in the Precambrian Atmosphere: An Evaluation of the Geological Evidence," *Geology* 10 (March 1982): 141.
7. Carl Sagan, "Ultraviolet Selection Pressure on the Earliest Organisms," *Journal of Theoretical Biology* 39 (April 1973): 195, 197.
8. Michael Denton, *Evolution: A Theory in Crisis* (Bethesda, Md.: Adler and Adler, 1986), 260–61.
9. Philip H. Abelson, "Chemical Events on the Primitive Earth," *Proceedings of the National Academy of Sciences* 55 (June 1966): 1365.
10. Stanley L. Miller and Leslie. E. Orgel, *The Origins of Life on the Earth* (Englewood Cliffs, N.J.: Prentice-Hall, 1974), 127.
11. Denton, 234.
12. Michael Behe, *Darwin's Black Box: The Biochemical Challenge to Evolution* (New York: The Free Press, 1996), 169–70; A. E. Wilder-Smith, *The Natural Sciences Know Nothing of Evolution* (Costa Mesa, Calif.: T.W.F.T. Publishers, 1981), 15–16.
13. Wilder-Smith, 16.
14. Behe, 170.
15. Francis Crick, *Life Itself: Its Origin and Nature* (New York: Simon and Schuster, 1981), 51–52.
16. Fred Hoyle, "The Big Bang in Astronomy," *New Scientist* 92 (19 November 1981): 527.
17. Bernard Lovell, *In the Centre of Immensities* (New York: Harper and Row, 1978), 63.
18. Harold G. Coffin, *Origin by Design* (Hagerstown, Md.: Review and Herald Publishing Association, 1983), 376.
19. Ibid.
20. Duane T. Gish, "The Origin of Life: Theories on the Origin of Biological Order," *Impact* 37 (July 1976): 3.
21. John Keosian, *Origin of Life* (1978), quoted in Duane T. Gish, *Creation Scientists Answer Their Critics* (El Cajon, Calif.: Institute for Creation Research, 1993), 373–74.
22. D. E. Green and R. F. Goldberger, *Molecular Insights into the Living Process* (New York: Academic Press, 1967), 406–7.
23. Denton, 329.
24. Carl Sagan, "Life," *Encyclopaedia Britannica*, 15th ed., vol. 22, 987.
25. Richard Dawkins, *The Blind Watchmaker* (New York: W. W. Norton, 1986), 17–18.
26. Denton, 328–29.
27. John Horgan, "In the Beginning," *Scientific American* 264 (February 1991): 119.
28. Denton, 234, 239.
29. Wilder-Smith, 86–90.

30. Caryl P. Haskins, "Advances and Challenges in Science in 1970," *American Scientist* 59 (May/June 1971): 305.
31. Robert Augros and George Stanciu, *The New Biology* (Boston: New Science Library, 1987), 191.
32. "Hoyle on Evolution," *Nature* 294 (12 November 1981): 105.
33. Crick, 88.
34. G. K. Chesterton, quoted in P. E. Hodgson, review of *Chesterton: A Seer of Science*, by Stanley L. Jaki, *National Review*, 5 June 1987, 47.
35. Loren Eiseley, *The Immense Journey* (New York: Random House, 1957), 199.
36. Robert W. Clark, *The Life of Ernst Chain: Penicillin and Beyond* (New York: St. Martin's Press, 1985), 148.

Chapter Eight. An Ape-man for All Seasons

1. Austin H. Clark, *The New Evolution: Zoogenesis* (Baltimore: Williams and Wilkins, 1930), 224–26.
2. Jerold M. Lowenstein and Adrienne L. Zihlman, "The Invisible Ape," *New Scientist* 120 (3 December 1988): 56.
3. John Reader, "Whatever Happened to Zinjanthropus?" *New Scientist* 89 (26 March 1981): 802.
4. Solly Zuckerman, *Beyond the Ivory Tower: The Frontiers of Public and Private Service* (New York: Taplinger, 1970), 64.
5. Pierre Teilhard de Chardin, *The Phenomenon of Man* (New York: Harper and Row, 1955), 219.
6. "Darwin Theory Proved True," *New York Times,* 22 December 1912, p. C1.
7. Henry Fairfield Osborn, *Evolution and Religion in Education* (New York: Charles Scribner's Sons, 1926), 103.
8. Henry Fairfield Osborn, "Hesperopithecus, the First Anthropoid Primate Found in America," *Science* 55 (5 May 1922): 464.
9. Harris Hawthorne Wilder, *The Pedigree of the Human Race* (New York: Henry Holt, 1926), 157.
10. Malcolm Bowden, *Ape-men: Fact or Fallacy?* (Bromley, England: Sovereign, 1977), 46–47.
11. Richard Carrington, *A Million Years of Man: The Story of Human Development as a Part of Nature* (New York: New American Library, 1964), 85.
12. Herbert Wendt, *From Ape to Adam* (New York: Bobbs-Merrill, 1972), 167.
13. Ibid., 168.
14. Marvin L. Lubenow, *Bones of Contention* (Grand Rapids, Mich.: Baker Books, 1992), 90.
15. G. H. R. von Koenigswald, *Meeting Prehistoric Man*, trans. Micheal [*sic*] Bullock (New York: Harper and Brothers, 1956), 38.
16. Ibid., 32.
17. Lubenow, 114–19.
18. Ian T. Taylor, *In the Minds of Men: Darwin and the New World Order* (Minneapolis: TFE Publishing, 1991), 224.
19. Arthur Keith, *The Antiquity of Man*, rev. ed., vol. 2 (Philadelphia: J. B. Lippincott, 1925), 440.
20. Stephen Jay Gould, *Eight Little Piggies* (New York: W. W. Norton, 1993), 135.

21. Franz Weidenreich, *Apes, Giants and Man* (Chicago: University of Chicago Press, 1946), 61.
22. William R. Fix, *The Bone Peddlers: Selling Evolution* (New York: Macmillan, 1984), 24.
23. Marcellin Boule and H. V. Vallois, *Fossil Men* (New York: Dryden Press, 1957), 141.
24. Carl O. Dunbar and Karl M. Waage, *Historical Geology*, 3rd ed. (New York: John Wiley and Sons, 1969), 499.
25. G. Elliot Smith, "The Discovery of Primitive Man in China," *Antiquity* 5 (March 1931): 36.
26. Henri Breuil, "Le feu et l'industrie de pierre et d'os dans le gisement du 'Sinanthropus' à Chou K'ou Tien," *L'Anthropologie* 42 (1932): 14.
27. Boule and Vallois, 145.
28. Duane T. Gish, *Evolution: The Fossils Still Say No!* (El Cajon, Calif.: Institute for Creation Research, 1995), 288.
29. Von Koenigswald, 51.
30. Earnest Albert Hooton, *Up from the Ape* (New York: Macmillan, 1931), 332.
31. Lubenow, 16.
32. William J. Straus and A. J. E. Cave, "Pathology and the Posture of Neanderthal Man," *Quarterly Review of Biology* 32 (December 1957): 358–59.
33. George Constable, *The Neanderthals* (New York: Time-Life Books, 1973), 101.
34. "Neandertal Noisemaker," *Science News* 150 (23 November 1996): 328.
35. *Nature* 77 (23 April 1908): 587.
36. Francis Ivanhoe, "Was Virchow Right about Neandertal?" *Nature* 227 (8 August 1970): 577–79.

Chapter Nine. The Reigning World Chimp

1. Jerold M. Lowenstein and Adrienne L. Zihlman, "The Invisible Ape," *New Scientist* 120 (3 December 1988): 59.
2. Ian T. Taylor, *In the Minds of Men: Darwin and the New World Order* (Minneapolis: TFE Publishing, 1991), 242.
3. T. C. Partridge, "Geomorphological Dating of Cave Openings at Makapansgat, Sterkfontein, Swartkrans and Taung," *Nature* 246 (9 November 1973): 75–79.
4. Reiner Protsch, "Age and Stratigraphic Position of Olduvai Hominid I," *Journal of Human Evolution* 3 (September 1974): 384.
5. Tim D. White, "Evolutionary Implications of Pliocene Hominid Footprints," *Science* 208 (11 April 1980): 175.
6. R. H. Tuttle and D. M. Webb, "The Pattern of Little Feet," *American Journal of Physical Anthropology* 78 (February 1989): 316.
7. Duane T. Gish, *Evolution: The Fossils Still Say No!* (El Cajon, Calif.: Institute for Creation Research, 1995), 276.
8. Richard E. Leakey, "Skull 1470," *National Geographic* 143 (June 1973): 819.
9. "Fossils Put Man at Four Million Years Old," *The Times* (London), 26 October 1974, p. 6.
10. Jack T. Stern, Jr. and Randall L. Susman, "The Locomotor Anatomy of Australopithecus Afarensis," *American Journal of Physical Anthropology* 60 (March 1983): 308.
11. Ibid., 298–99.

12. Solly Zuckerman, *Beyond the Ivory Tower: The Frontiers of Public and Private Service* (New York: Taplinger, 1970), 93.
13. Stephen Jay Gould, "A Short Way to Big Ends," *Natural History* 95 (January 1986): 28.
14. Charles Oxnard, *The Order of Man: A Biomathematical Anatomy of the Primates* (New Haven: Yale University Press, 1984), 332.
15. Charles Oxnard, *Fossils, Teeth and Sex: New Perspectives on Human Evolution* (Seattle: University of Washington Press, 1987), 227.
16. Gish, *The Fossils Still Say No!* 256.
17. Pat Shipman, "These Ears Were Made for Walking," *New Scientist* 143 (30 July 1994): 28–29.
18. "The Case for a Living Link," *Time*, 4 December 1978, 82.
19. Wray Herbert, "Lucy's Uncommon Forbear," *Science News* 123 (5 February 1983): 92.
20. Henry M. McHenry, "Fossils and the Mosaic Nature of Human Evolution," *Science* 190 (31 October 1975): 428.
21. Arthur Keith, *The Antiquity of Man* (London: Williams and Norgate, 1920), 283–85.
22. Ibid., 245.
23. Earnest Albert Hooton, *Apes, Men and Morons* (New York: G. P. Putnam's Sons, 1937), 107.
24. Robert Broom and G. W. H. Schepers, *The South African Fossil Ape-Men, The Australopithecinae* (Pretoria: Transvaal Museum, 1946), 257.
25. Michael Denton, *Evolution: A Theory in Crisis* (Bethesda, Md.: Adler and Adler, 1986), 177.
26. Stephen Molnar, *Races, Types, and Ethnic Groups: The Problem of Human Variation* (Englewood Cliffs, N.J.: Prentice-Hall, 1975), 57.
27. Roger Lewin, *Bones of Contention: Controversies in the Search for Human Origins* (New York: Simon and Schuster, 1987), 27.
28. Greg Kirby, address at meeting of Biology Teachers' Association, South Australia, 1976, quoted in *The Revised Quote Book* (Acacia Ridge, Queensland, Australia: Creation Science Foundation, 1990), 16.
29. Ian Anderson, "Hominoid Collarbone Exposed as Dolphin's Rib," *New Scientist* 98 (28 April 1983), 199.
30. Marvin L. Lubenow, *Bones of Contention* (Grand Rapids, Mich.: Baker Books, 1992), 253.
31. Lewin, 26.
32. Ibid., 24.
33. *The Weekend Australian*, 7–8 May 1983, Magazine, 3, quoted in *The Revised Quote Book*, 14.
34. David J. Jefferson, "This Anthropologist Has a Style That Is Bone of Contention," *Wall Street Journal*, 31 January 1995, pp. A1, A12.
35. J. S. Jones, "A Thousand and One Eves," *Nature* 345 (31 May 1990): 395.
36. Christopher B. Stringer, "The Legacy of Homo Sapiens," *Scientific American* 268 (May 1993): 138.
37. "Three Human Species Coexisted Eons Ago, New Data Suggest," *New York Times*, 13 December 1996, p. A1.
38. Robert Martin, "Man is Not an Onion," *New Scientist* 75 (4 August 1977): 283.
39. Lowenstein and Zihlman, 58.
40. "Whose Ape Is It, Anyway?" *Science News* 125 (9 June 1984): 361.
41. Lyall Watson, "The Water People," *Science Digest* 90 (May 1982): 44.

42. John Reader, "Whatever Happened to Zinjanthropus?" *New Scientist* 89 (26 March 1981): 802.
43. John W. Klotz, "The Case for Evolution," in *Darwin, Evolution and Creation*, ed. Paul A. Zimmerman, (St. Louis: Concordia, 1959), 128.
44. John Whitcomb, *The Early Earth* (Grand Rapids, Mich.: Baker Books, 1986), 130.
45. Noam Chomsky, *Language and Mind* (New York: Harcourt, Brace, Jovanovich, 1972), 67.

Chapter Ten. Old Myths Never Die–They Only Fade Away

1. Ernst Haeckel, *The Riddle of the Universe at the Close of the Nineteenth Century*, trans. Joseph McCabe (New York: Harper and Brothers, 1900), 65–66.
2. Francis Hitching, *The Neck of the Giraffe* (New York: Ticknor and Fields, 1982), 202–4.
3. Michael Pitman, *Adam and Evolution* (London: Rider, 1984), 120.
4. J. Assmuth and Ernest R. Hull, *Haeckel's Frauds and Forgeries* (Bombay: Examiner Press, 1915), 26.
5. Ibid., 24.
6. Ibid., 29.
7. Ibid., 30.
8. Jan Langman, *Medical Embryology*, 3rd ed. (Baltimore: Williams and Wilkins, 1975), 260–62.
9. Keith Stewart Thomson, "Ontogeny and Phylogeny Recapitulated," *American Scientist* 76 (May/June 1988): 273.
10. Sabine Schwabenthan, "Life Before Birth," *Parents* (October 1979): 50.
11. "An Embryonic Liar," *The Times* (London), 11 August 1997, p. 14.
12. Walter J. Bock, "Evolution by Orderly Law," *Science* 164 (9 May 1969): 684.
13. A. Rendle Short, "Some Recent Literature Concerning the Origin of Man," *Journal of the Transactions of the Victoria Institute* 67 (1935): 256.
14. Stephen Jay Gould, "Dr. Down's Syndrome," *Natural History* 89 (April 1980): 144.
15. Gavin de Beer, "Darwin and Embryology," in *A Century of Darwin*, ed. Samuel Anthony Barnett (Freeport, N.Y.: Books for Libraries Press, 1969), 159.
16. *The World's Most Famous Court Trial: Tennessee Evolution Case* (Dayton, Tenn.: Bryan College, 1990), 268.
17. Charles Darwin, *The Origin of Species* (1872; reprint, New York: Random House, 1993), 626.
18. Richard Goldschmidt, *The Material Basis of Evolution* (New Haven: Yale University Press, 1940), 390.
19. Ibid., 395.
20. Sewall Wright, "Character Change, Speciation, and the Higher Taxa," *Evolution* 36 (May 1982): 440.
21. Stephen Jay Gould, "The Return of Hopeful Monsters," *Natural History* 86 (June/July 1977): 22.
22. Niles Eldredge and Stephen Jay Gould, "Punctuated Equilibria: an Alternative to Phyletic Gradualism," in *Models in Paleobiology,* ed. Thomas J. M. Schopf (San Francisco: Freeman Cooper, 1972), 82–115.
23. Robert E. Ricklefs, "Paleontologists Confronting Macroevolution," *Science* 199 (6 January 1978): 59.

24. Kenneth J. Hsu, "Sedimentary Petrology and Biologic Evolution," *Journal of Sedimentary Petrology* 56 (September 1986): 730.
25. Søren Løvtrup, *Darwinism: The Refutation of a Myth* (New York: Croom Helm, 1987), 422.
26. William Dawson, *The Story of Earth and Man* (New York: Harper and Brothers, 1887), 317, 322, 330, 339.
27. Henry M. Morris, *Men of Science—Men of God* (El Cajon, Calif.: Master Books, 1988), 84.
28. N. J. Mitchell, *Evolution and the Emperor's New Clothes* (United Kingdom: Roydon Publications, 1983), title page, quoted in *The Revised Quote Book* (Acacia Ridge, Queensland, Australia: Creation Science Foundation, 1990), 5.
29. John Fred Meldau, ed., *Witnesses Against Evolution* (Denver: Christian Victory Publishing, 1968), 13.
30. Albert Fleischmann, "The Doctrine of Organic Evolution in the Light of Modern Research," *Journal of the Transactions of the Victoria Institute* 65 (1933): 194–95, 205–6, 208–9.
31. *The Advocate*, 8 March 1984, 17, quoted in *The Revised Quote Book*, 5.
32. Wolfgang Smith, *Teilhardism and the New Religion* (Rockford., Ill.: Tan Books, 1988), 5–6.
33. Phillip E. Johnson, *Darwin on Trial* (Washington, D.C.: Regnery Gateway, 1991), 10, 157.
34. Colin Patterson, address at American Museum of Natural History, 5 November 1981, quoted in *The Revised Quote Book*, 4.

Chapter Eleven. The Big Bang Goes Blooey

1. Paul Davies, *The Edge of Infinity* (New York: Simon and Schuster, 1981), 161.
2. Cicero, *On the Nature of the Gods*, trans. Horace McGregor (Harmondsworth, England: Penguin Books, 1972), 144–45.
3. Robert Kirshner, "The Universe as a Lattice," *Nature* 385 (9 January 1997): 112.
4. Ben Patrusky, "Why is the Cosmos 'Lumpy'?" *Science 81* (June 1981): 96.
5. Eric J. Lerner, "The Big Bang Never Happened," *Discover* 9 (June 1988): 78.
6. Fred Hoyle, "The Big Bang Under Attack," *Science Digest* 92 (May 1984): 84.
7. Paul Kalas, "Dusky Disks and Planet Mania," *Science* 281 (10 July 1998): 182.
8. John Hudson Tiner, *Isaac Newton: Inventor, Scientist, and Teacher* (Milford, Mich.: Mott Media, 1975), i.
9. Jerry Bergman, "The Earth: Unique in All the Universe," *Impact* 144 (June 1985): 2–3.
10. Stuart E. Nevins, "Planet Earth: Plan or Accident?" *Impact* 14 (May 1974): 4.
11. Ibid.
12. Fred Hoyle, "Where the Earth Came From," *Harper's*, March 1951, 65.
13. Jerry E. Bishop, "New Theories of Creation," *Science Digest* 72 (October 1972): 42.
14. Richard A. Kerr, "The Solar System's New Diversity," *Science* 265 (2 September 1994): 1360.
15. Charles J. Lada and Frank H. Shu, "The Formation of Sunlike Stars," *Science* 248 (4 May 1990): 572.
16. Marcus Chown, "Let There Be Light," *New Scientist* 157 (7 February 1998): 30.
17. Corey S. Powell, "A Matter of Timing," *Scientific American* 267 (October 1992): 30.
18. Arthur Eddington, *The Expanding Universe* (New York: Macmillan, 1933), 25.

19. Genesis 1:16.
20. Sharon Begley, "Science Finds God," *Newsweek,* 20 July 1998, 50.
21. James Jeans, *The Mysterious Universe* (New York: Macmillan, 1930), 144, 159.
22. Henry M. Morris, *Men of Science—Men of God* (El Cajon, Calif.: Master Books, 1988), 85.

Chapter Twelve. Earth, Dahling, You Don't Look a Day Over Five Billion

1. John Whitcomb, *The Early Earth* (Grand Rapids, Mich.: Baker Books, 1986), 13.
2. Gloria B. Lubkin, "Analyses of Historical Data Suggest Sun is Shrinking," *Physics Today* 32 (September 1979): 17.
3. Jonathan Sarfati, "The Moon: The Light that Rules the Night," *Creation Ex Nihilo* 20 (September-November 1998): 38.
4. Fred Hoyle, *Frontiers of Astronomy* (New York: Harper and Brothers, 1955), 11.
5. F. A. Paneth, "The Frequency of Meteorite Falls Throughout the Ages," in *Vistas in Astronomy*, vol. 2, ed. Arthur Beer (New York: Pergamon Press, 1956), 1681.
6. W. A. Tarr, "Meteorites in Sedimentary Rocks?" *Science* 75 (1 January 1932): 17.
7. Melvin A. Cook, "Where is the Earth's Radiogenic Helium?" *Nature* 179 (26 January 1957): 213.
8. "What Happened to the Earth's Helium?" *New Scientist* 24 (3 December 1964): 631–32.
9. Larry Vardiman, *The Age of the Earth's Atmosphere: A Study of the Helium Flux through the Atmosphere* (El Cajon, Calif.: Institute for Creation Research, 1990).
10. Thomas G. Barnes, "Physics: A Challenge to 'Geologic Time,'" *Impact* 16 (July 1974): 2–3.
11. Keith L. McDonald and Robert H. Gunst, "An Analysis of the Earth's Magnetic Field from 1835 to 1965," ESSA Technical Report IER 1 (Washington, D.C.: U.S. Government Printing Office, July 1967), Table 3, p. 15, cited in Thomas G. Barnes, "Young Age for Moon and Earth," *Impact* 110 (August 1982): 3.
12. "Magsat Down, Magnetic Field Declining," *Science News* 117 (28 June 1980): 407.
13. Jeremy Bloxham and David Gubbins, "The Evolution of the Earth's Magnetic Field," *Scientific American* 261 (December 1989): 71.
14. Russell Humphreys, "The Earth's Magnetic Field is Young," *Impact* 242 (August 1993): 3–4.
15. A. A. Humphreys and H. L. Abbot, *Report upon the Physics and Hydraulics of the Mississippi River* (Philadelphia: J. B. Lippincott, 1861), 425.
16. Henry M. Morris, *Scientific Creationism* (Green Forest, Ark.: Master Books, 1985), 167–69.
17. Henry M. Morris and John D. Morris, *Society and Creation* (Green Forest, Ark.: Master Books, 1996), 147.
18. Stuart E. Nevins, "Evolution: The Oceans Say No!" *Impact* 8 (1973): 3.
19. John D. Morris, *The Young Earth* (Green Forest, Ark.: Master Books, 1994), 89.
20. Ibid., 85–87.
21. Henry M. Morris and Gary E. Parker, *What Is Creation Science?* (El Cajon, Calif.: Master Books, 1987), 283–84.
22. H. S. Lipson, "A Physicist Looks at Evolution," *Physics Bulletin* 31 (May 1980): 138.

Chapter Thirteen. Assumptions Aplenty

1. Elizabeth K. Ralph and Henry M. Michael, "Twenty-five Years of Radiocarbon Dating," *American Scientist* 62 (September/October 1974): 555.
2. Robert Stuckenrath, "Radiocarbon: Some Notes from Merlin's Diary," *Annals of the New York Academy of Science* 288 (1977): 186.
3. Harold G. Coffin, *Origin by Design* (Hagerstown, Md.: Review and Herald Publishing Association, 1983), 315–17.
4. Larry Vardiman, "The Sky Has Fallen," *Impact* 128 (February 1984): 2.
5. Ernst Antevs, "Geological Tests of the Varve and Radiocarbon Chronologies," *Journal of Geology* 65 (March 1957): 129–30.
6. Alan C. Riggs, "Major Carbon-14 Deficiency in Modern Snail Shells from Southern Nevada Springs," *Science* 224 (6 April 1984): 58.
7. Wakefield Dort, Jr., "Mummified Seals of Southern Victoria Land," *Antarctic Journal of the United States* 6 (September-October 1971): 211.
8. Willard F. Libby, "Radiocarbon Dating," *American Scientist* 44 (January 1956): 107.
9. J. Gordon Ogden, III, "The Use and Abuse of Radiocarbon Dating," *Annals of the New York Academy of Science* 288 (1977): 173.
10. Robert E. Lee, "Radiocarbon: Ages in Error," *Anthropological Journal of Canada* 19, no. 3 (1981): 9, 26–27.
11. Stuckenrath, 188.
12. John D. Morris, *The Young Earth* (Green Forest, Ark.: Master Books, 1994), 91.
13. A. E. Wilder-Smith, *The Natural Sciences Know Nothing of Evolution* (Costa Mesa, Calif.: T.W.F.T. Publishers, 1981), 133.
14. *McLean vs. Arkansas State Board of Education*, Testimony of Gary B. Dalrymple, Court Reporter S. Smith, 449–50 (Little Rock, U.S. District Court, 1981), quoted in Robert V. Gentry, *Creation's Tiny Mystery* (Knoxville, Tenn.: Earth Science Associates, 1992), 112–13.
15. Frederic B. Jueneman, "Secular Catastrophism," *Industrial Research and Development* (June 1982): 21.
16. William D. Stansfield, *The Science of Evolution* (New York: Macmillan, 1977), 84.
17. Steven A. Austin, ed., *Grand Canyon: Monument to Disaster* (Santee, Calif.: Institute for Creation Research, 1994), 120–25.
18. John G. Funkhouser and John J. Naughton, "Radiogenic Helium and Argon in Ultramafic Inclusions from Hawaii," *Journal of Geophysical Research* 73 (July 1968): 4601–7.
19. G. B. Dalrymple, "40 Ar/36 Ar Analyses of Historical Lava Flows," *Earth and Planetary Science Letters* 6 (1969): 47–55.
20. Ian McDougall, H. A. Polach and J. J. Stipp, "Excess Radiogenic Argon in Young Subaerial Basalts from Auckland Volcanic Field, New Zealand," *Geochimica et Cosmochimica Acta* 33 (December 1969): 1485, 1499.
21. Henry M. Morris, "Radiometric Dating and the Bible: A Historical Review," *Back to Genesis* (August 1997): d.
22. A. Brock and G. Ll. Isaac, "Paleomagnetic Stratigraphy and Chronology of Hominid-bearing Sediments East of Lake Rudolf, Kenya," *Nature* 247 (8 February 1974): 347.
23. Marvin L. Lubenow, *Bones of Contention* (Grand Rapids, Mich.: Baker Books, 1992), 253–66.
24. A. Hayatsu, "K-Ar Isochron Age of the North Mountain Basalt, Nova Scotia," *Canadian Journal of Earth Sciences* 16 (April 1979): 974.

25. J. F. Evernden and J. R. Richards, "Potassium-argon Ages in Eastern Australia," *Journal of the Geological Society of Australia* 9, part 1 (July 1962): 3.
26. Richard L. Mauger, "K-Ar Ages of Biotites from Tuffs in Eocene Rocks of the Green River, Washakie, and Uinta Basins, Utah, Wyoming, and Colorado," *Contributions to Geology* (University of Wyoming) 15 (Winter 1977): 37.
27. T. Säve-Söderbergh and Ingrid U. Olsson, "C14 Dating and Egyptian Chronology," *Radiocarbon Variations and Absolute Chronology*, ed. Ingrid U. Olsson (New York: John Wiley and Sons, 1970), 35.
28. Gentry, 82.
29. Ibid., 31.
30. R. G. Kazmann, "It's About Time: 4.5 Billion Years," *Geotimes* 23 (September 1978): 19.
31. Gary E. Parker, "From Evolution to Creation: A Personal Testimony," *Impact* 49 (July 1977): 3.

Chapter Fourteen. Rocks of Ages

1. George P. Fisher, *The Life of Benjamin Silliman*, vol. 2 (New York: Charles Scribner, 1866), 148.
2. Steven A. Austin, ed., *Grand Canyon: Monument to Disaster* (Santee, Calif.: Institute for Creation Research, 1994), 24.
3. Francis Hitching, *The Neck of the Giraffe* (New York: Ticknor and Fields, 1982), 227.
4. Ibid.
5. Gertrude Himmelfarb, *Darwin and the Darwinian Revolution* (Gloucester, Mass.: Peter Smith, 1967), 387.
6. Ronald R. West, "Paleoecology and Uniformitarianism," *Compass* 45 (May 1968): 216, quoted in *The Revised Quote Book* (Acacia Ridge, Queensland, Australia: Creation Science Foundation, 1990), 10.
7. Niles Eldredge, *Time Frames: The Rethinking of Darwinian Evolution and the Theory of Punctuated Equilibria* (New York: Simon and Schuster, 1985), 52.
8. Tom Kemp, "A Fresh Look at the Fossil Record," *New Scientist* 108 (5 December 1985): 66.
9. R. H. Rastall, "Geology," *Encyclopaedia Britannica* (1956), vol. 10, 168, quoted in *The Revised Quote Book*, 25.
10. J. E. O'Rourke, "Pragmatism versus Materialism in Stratigraphy," *American Journal of Science* 276 (January 1976): 47.
11. Steven A. Austin, "Ten Misconceptions about the Geologic Column," *Impact* 137 (November 1984): 2.
12. Henry M. Morris and Gary E. Parker, *What Is Creation Science?* (El Cajon, Calif.: Master Books, 1987), 163.
13. "Mountain Building in the Mediterranean," *Science News* 98 (17 October 1970): 316.
14. John C. Whitcomb and Henry M. Morris, *The Genesis Flood: The Biblical Record and Its Scientific Implications* (Phillipsburg, N.J.: Presbyterian and Reformed Publishing, 1961), 181.
15. Austin, "Ten Misconceptions," 2.
16. Austin, *Grand Canyon*, 147.
17. "Striking Oil in the Laboratory," *Science News* 125 (24 March 1984): 187.

18. Elizabeth Pennisi, "Water, Water Everywhere," *Science News* 143 (20 February 1993): 124.
19. "Basic Coal Studies Refute Current Theories of Formation," *Research and Development* (February 1984): 92.
20. Anna K. Behrensmeyer, "Taphonomy and the Fossil Record," *American Scientist* 72 (November/December 1984): 560.
21. Norman D. Newell, "Symposium on Fifty Years of Paleontology: Adequacy of the Fossil Record," *Journal of Paleontology* 33 (May 1959): 492.
22. Harry S. Ladd, "Ecology, Paleontology, and Stratigraphy," *Science* 129 (9 January 1959): 72.
23. Hugh Miller, *The Old Red Sandstone* (1857; reprint, New York: Arno Press, 1978), 221–22.
24. Harold G. Coffin, *Origin by Design* (Hagerstown, Md.: Review and Herald Publishing Association, 1983), 81.
25. Derek V. Ager, *The Nature of the Stratigraphic Record*, 3rd. ed. (New York: John Wiley and Sons, 1993), 67–68.
26. J. Harlen Bretz, "The Lake Missoula Floods and the Channeled Scabland," *Journal of Geology* 77 (September 1969): 541.
27. Sigurdur Thorarinsson, *Surtsey, the New Island in the North Atlantic* (New York: Viking, 1967).
28. John D. Morris, *The Young Earth* (Green Forest, Ark.: Master Books, 1994), 107.
29. Steven A. Austin, "Mount St. Helens and Catastrophism," *Impact* 157 (July 1986): 1–2.
30. R. J. Rice, "The Canyon Conundrum," *Geographical Magazine* 55 (May 1983): 292.
31. Earle E. Spamer, "The Development of Geological Studies in the Grand Canyon," *Tryonia* 17 (June 1989): 39.
32. Austin, *Grand Canyon*, 92.
33. Ibid.
34. Edgar B. Heylmun, "Should We Teach Uniformitarianism?" *Journal of Geological Education* 19 (January 1971): 35.
35. Stephen Jay Gould, "Catastrophes and Steady State Earth," *Natural History* 84 (February 1975): 15–17.
36. Warren D. Allmon, "Post-Gradualism," review of *The New Catastrophism,* by Derek V. Ager, *Science* 262 (1 October 1993): 122.
37. Ager, 80.

Chapter Fifteen. The Flood Remembered

1. For a complete list, see Tim F. LaHaye and John D. Morris, *The Ark on Ararat* (Nashville: Thomas Nelson, 1976), 233–36.
2. Henry Rowe Schoolcraft, *History of the Indian Tribes of the United States* (Philadelphia: J. B. Lippincott, 1857), 571.
3. The Bible recounts the Flood in Genesis 6–8.
4. LaHaye and Morris, 237.
5. Duane T. Gish, *Dinosaurs by Design* (Green Forest, Ark.: Master books, 1992), 74.
6. Ibid.
7. "Genesis According to the Miao People," *Impact* 214 (April 1991): 2.
8. LaHaye and Morris, 237.

9. Lee Spetner, *Not By Chance!: Shattering the Modern Theory of Evolution* (Brooklyn, N.Y.: Judaica Press, 1997), 204.
10. William J. Robbins and Harold W. Rickett, *Botany* (New York: D. Van Nostrand, 1939), 559.
11. John C. Whitcomb and Henry M. Morris, *The Genesis Flood: The Biblical Record and Its Scientific Implications* (Phillipsburg, N.J.: Presbyterian and Reformed Publishing, 1961), 67–68.
12. Genesis 7:11.
13. Psalms 104:8 (New American Standard Bible).
14. Harold G. Coffin, *Origin by Design* (Hagerstown, Md.: Review and Herald Publishing Association, 1983), 175.
15. Tage Nillson, *The Pleistocene: Geology and Life in the Quaternary Ice Age* (Boston: Reidel, 1983), 111, 223.
16. Coffin, 261.
17. Larry Vardiman, "The Sky Has Fallen," *Impact* 128 (February 1984): 2.
18. Michael J. Novacek et al., "Fossils of the Flaming Cliffs," *Scientific American* 271 (December 1994): 60, 62.
19. Henry M. Morris and John D. Morris, *Science and Creation* (Green Forest, Ark.: Master Books, 1996), 237.
20. John Woodmorappe, *Noah's Ark: A Feasibility Study* (Santee, Calif.: Institute for Creation Research, 1996), 113–17.
21. William F. Albright, "Recent Discoveries in Bible Lands," in *Analytical Concordance to the Bible*, by Robert Young (New York: Funk and Wagnalls, 1955), 30.
22. Bill Cooper, *After the Flood* (Chichester, England: New Wine Press, 1995), 41.
23. Ibid., 211.
24. Ibid., 83–96.
25. Kenneth Sisam, *Anglo-Saxon Royal Genealogies* (London: Oxford University Press, 1954), 320.
26. Cooper, 97–106.
27. Jonathan Barnes, *Early Greek Philosophy* (London: Penguin Books, 1987), 95–97.
28. John Ross, *The Original Religion of China* (London: Oliphant, Anderson and Ferrier, 1909), 295.
29. James Legge, *The Notions of the Chinese Concerning God and Spirits* (Hong Kong: Hong Kong Register Office, 1852), 28.
30. Ibid.
31. Ross, 296.
32. Richard A. Kerr, "Pathfinder Strikes a Rocky Bonanza," *Science* 277 (11 July 1997): 173.

Chapter Sixteen. Dinosaurs, Dragons and Ice

1. Peter Dickinson, *The Flight of Dragons* (New York: Harper and Row, 1979), 127.
2. Rhoda Blumberg, *The Truth about Dragons* (New York: Four Winds Press, 1980), 8.
3. *The Enchanted World: Dragons* (Alexandria, Va.: Time-Life Books, 1984), 23.
4. "Bushmen's Paintings Baffling to Scientists," *Los Angeles Herald Examiner*, 7 January 1970, quoted in Henry M. Morris, *That Their Words May Be Used Against Them* (San Diego: Institute for Creation Research, 1997), 252.
5. Ken Ham, Andrew Snelling and Carl Wieland, *The Answers Book* (El Cajon, Calif.: Master Books, 1991), 33.

6. Dickinson, 79.
7. *The Enchanted World: Dragons*, 33.
8. Ibid., 79.
9. Ibid., 57.
10. D. Leland Niermann, "Dinosaurs and Dragons," *Creation Ex Nihilo Technical Journal* 8, no. 1 (1994): 100.
11. Blumberg, 52.
12. Geoffrey of Monmouth, *The History of the Kings of Britain*, trans. Sebastian Evans, rev. Charles W. Dunn (New York: Dutton, 1958), 60.
13. Duane T. Gish, *Dinosaurs by Design* (Green Forest, Ark.: Master Books, 1992), 40, 82.
14. Bill Cooper, *After the Flood* (Chichester, England: New Wine Press, 1995), 133–34.
15. Ibid., 133.
16. Ibid., 153–54.
17. Ibid., 156.
18. Brad Steiger, *Worlds Before Our Own*, (New York: G. P. Putnam's Sons, 1978), 42–43.
19. P. A. Armstrong, *The Piasa or the Devil Among the Indians* (1887), quoted in Steiger, 42.
20. Steiger, 47–48.
21. *The Enchanted World: Dragons*, 79.
22. John Sinclair, *The Statistical Account of Scotland*, vol. 6 (Edinburgh: William Creech, 1793), 467.
23. "Very Like a Whale," *The Illustrated London News*, 9 February 1856, p. 166.
24. "A Strange Winged Monster Discovered and Killed on the Huachuca Desert," *Tombstone Epitaph*, 26 April 1890, p. 3.
25. Job 41:20–27.
26. *The Travels of Marco Polo* (New York: Dutton, 1908), 315, 264.
27. Michael J. Oard, *An Ice Age Caused by the Genesis Flood* (El Cajon, Calif.: Institute for Creation Research, 1990), 46.
28. Ibid., 148.
29. N. J. Shackleton and J. P. Kennett, "Paleotemperature History of the Cenozoic and the Initiation of Antarctic Glaciation: Oxygen and Carbon Isotope Analyses in DSDP Sites 277, 279, 281," *Initial Reports of the Deep Sea Drilling Project* 29 (Washington, D.C.: U. S. Government Printing Office, 1975), 743–55.
30. Larry Vardiman, "Out of Whose Womb Came the Ice?" *Impact* 254 (August 1994): 3.
31. F. Barrows Colton, "Weather Works and Fights for Man," *National Geographic* 84 (December 1943): 668.
32. Exodus 3:8; Deuteronomy 8:7–8.
33. Oard, 78–80.
34. Steven Petrow, "The Lost Squadron," *Life*, December 1992, 65.

Chapter Seventeen. Trial by Hollywood

1. Jerome Lawrence and Robert E. Lee, *Inherit the Wind* (1955; reprint, New York: Bantam, 1960), page preceding Act One.
2. L. Sprague de Camp, *The Great Monkey Trial* (Garden City, N.Y.: Doubleday, 1968), 432.

3. John T. Scopes and James Presley, *Center of the Storm* (New York: Holt, Rinehart and Winston, 1967), 60.
4. Ibid., 187–88.
5. Ibid., 134.
6. Ibid., 60.
7. Edward J. Larson, *Summer for the Gods: The Scopes Trial and America's Continuing Debate Over Science and Religion* (New York: Basic Books, 1997), 96.
8. *The World's Most Famous Court Trial: Tennessee Evolution Case* (Dayton, Tenn.: Bryan College, 1990) (hereinafter referred to as "Transcript"), 226.
9. Transcript, 322.
10. William Jennings Bryan, "Mr. Bryan on Evolution," *Reader's Digest*, August 1925, 213–14.
11. Transcript, 175–76.
12. Transcript, 117.
13. Transcript, 133–43.
14. Transcript, 205–6.
15. Larson, 184.
16. Transcript, 206–7; Scopes and Presley, 160.
17. Transcript, 306.
18. Larson, 182–83.
19. "Bryan Now Regrets Barring of Experts," *New York Times*, 18 July 1925, p. 2.
20. Transcript, 299.
21. Transcript, 284.
22. Transcript, 288.
23. Transcript, 285.
24. Transcript, 296.
25. Transcript, 287.
26. Transcript, 288.
27. Transcript, 293.
28. Kevin Tierney, *Darrow: A Biography* (New York: Thomas Y. Crowell, 1979), 369.
29. Clarence Darrow, *The Story of My Life* (New York: Charles Scribner's Sons, 1932), 259–60.
30. Transcript, 308.
31. De Camp, 62–63.
32. Ibid., 127.
33. Scopes and Presley, 206–7.
34. Larson, 200.
35. "The New Pictures," *Time*, 17 October 1960, 95.
36. Carol Iannone, "The Truth About Inherit the Wind," *First Things* 70 (February 1997).
37. "Mixed Bag," *The New Yorker*, 30 April 1955, 67.
38. Larson, 242.

Chapter Eighteen. Have You Murdered Anybody Since Breakfast?

1. John C. Burnham, "A Discarded Consensus," review of *Outcasts from Evolution: Scientific Attitudes of Racial Inferiority, 1859–1900*, by John S. Haller, Jr., *Science* 175 (4 February 1972): 506.

2. Charles Darwin, *Life and Letters*, ed. Francis Darwin, vol. 1 (1888; reprint, New York: Johnson Reprint, 1969), 316.
3. Thomas Huxley, *Lay Sermons, Addresses and Reviews* (New York: Appleton, 1870), 20.
4. Ernst Haeckel, *The Wonders of Life* (New York: Harper, 1904), 56–57.
5. Henry Fairfield Osborn, "The Evolution of Human Races," *Natural History* (January/February 1926), reprinted in *Natural History* 89 (April 1980): 129.
6. Ibid.
7. Edwin G. Conklin, *The Direction of Human Evolution* (New York: Scribner's, 1923), 34, 46, 53.
8. George William Hunter, *A Civic Biology: Presented in Problems* (New York: American, 1914), 196.
9. Stephen Jay Gould, *Ever Since Darwin: Reflections in Natural History* (New York: W. W. Norton, 1977), 217.
10. Carl Wieland, "The Lies of Lynchburg," *Creation Ex Nihilo* 19 (September-November 1997): 23.
11. Haeckel, 21, 119–20.
12. Ibid., 118–19.
13. *The World's Most Famous Court Trial: Tennessee Evolution Case* (Dayton, Tenn.: Bryan College, 1990), 337.
14. Gertrude Himmelfarb, *Darwin and the Darwinian Revolution* (Gloucester, Mass.: Peter Smith, 1967), 416–17.
15. Allan Chase, *The Legacy of Malthus* (New York: Alfred A. Knopf, 1977), 135.
16. Charles Darwin, *The Descent of Man and Selection in Relation to Sex* (New York: D. Appleton, 1896), 156.
17. Ibid., 133–34.
18. Jerry Bergman, "Eugenics and the Development of Nazi Race Policy," *Perspectives on Science and Christian Faith* 44 (June 1992): 109.
19. Arthur Keith, *Evolution and Ethics* (New York: Putnam, 1947), 230.
20. Ibid., 28.
21. George J. Stein, "Biological Science and the Roots of Nazism," *American Scientist* 76 (January/February 1988): 53.
22. Adolf Hitler, *Mein Kampf*, trans. Ralph Manheim (Boston: Houghton Mifflin, 1943), 285.
23. A. E. Wilder-Smith, *The Natural Sciences Know Nothing of Evolution* (Costa Mesa, Calif.: T.W.F.T. Publishers, 1981), 162.
24. Stein, 52.
25. Conway Zirkle, *Evolution, Marxian Biology, and the Social Scene* (Philadelphia: University of Philadelphia Press, 1959), 86.
26. Arthur Keith, *Darwin Revalued* (London: Watts, 1955), 233–34.
27. Emelian Yaroslavsky, *Landmarks in the Life of Stalin* (Moscow: Foreign Languages Publishing House, 1940), 8–9.
28. Stein, 52.
29. James Rachels, *Created from Animals: The Moral Implications of Darwinism* (New York: Oxford University Press, 1990), 64.
30. Andrew Carnegie, *Autobiography of Andrew Carnegie*, ed. John C. Van Dyke (1920; reprint, Boston: Northeastern University Press, 1986), 327.
31. D. R. Oldroyd, *Darwinian Impacts: An Introduction to the Darwinian Revolution* (Atlantic Highlands, N. J.: Humanities Press, 1980), 216.

32. Ian T. Taylor, *In the Minds of Men: Darwin and the New World Order* (Minneapolis: TFE Publishing, 1991), 372, 386.
33. Edward J. Larson, *Summer for the Gods: The Scopes Trial and America's Continuing Debate Over Science and Religion* (New York: Basic Books, 1997), 183.
34. Darwin, *Descent of Man*, 563–64.
35. Ibid., 564–65.
36. Stephen Jay Gould, *The Mismeasure of Man* (New York: W. W. Norton, 1981), 104–5.

Chapter Nineteen. The Boomers Doomed

1. *The World's Most Famous Court Trial: Tennessee Evolution Case* (Dayton, Tenn.: Bryan College, 1990), 179–80.
2. Ibid., 175.
3. E. O. Wilson, "Toward a Humanistic Biology," *The Humanist* (September/October 1982): 40.
4. Jeremy Rifkin, *Algeny* (New York: Viking, 1983), 244.
5. Julian Huxley, in *Issues in Evolution*, vol. 3, ed. Sol Tax (Chicago: University of Chicago Press, 1960), 45.
6. Michael Walker, "To Have Evolved or to Have Not? That is the Question," *Quadrant* 25 (October 1981): 45.
7. Aldous Huxley, *Ends and Means: An Inquiry into the Nature of Ideals and into the Methods Employed for Their Realization* (New York: Harper and Brothers, 1937), 312, 316.
8. Will Durant, "We Are in the Last Stage of a Pagan Period," *Chicago Tribune* (April 1980), quoted in Henry M. Morris and John D. Morris, *Society and Creation* (Green Forest, Ark.: Master Books, 1996), 80.
9. Edward E. Ericson, Jr., "Solzhenitsyn: Voice from the Gulag," *Eternity* (October 1985): 24.
10. James Crouse and Dale Trusheim, *The Case Against the SAT* (Chicago: University of Chicago Press, 1988), 134.
11. Edward J. Larson, *Summer for the Gods: The Scopes Trial and America's Continuing Debate Over Science and Religion* (New York: Basic Books, 1997), 230–31.
12. Julian Huxley, "The Evolutionary Vision," in *Issues in Evolution,* vol. 3, 252.
13. "News and Notes," *Movieguide* (July B 1996): 16.
14. Jefferson to Peter Carr, *The Papers of Thomas Jefferson*, vol. 8, ed. Julian P. Boyd (Princeton, N.J.: Princeton University Press, 1953), 407.
15. William J. Murray, *My Life Without God* (Eugene, Oreg.: Harvest House, 1992), 303.

Chapter Twenty. Good Company

1. Henry M. Morris and Gary E. Parker, *What Is Creation Science?* (El Cajon, Calif.: Master Books, 1987), 300.
2. Jane Ingraham, "Darwin is Dead," *The New American* 2 (24 November 1986): 34.
3. William F. Jasper, "Tennessee's Textbook Trial," *The New American* 2 (24 November 1986): 51.
4. "Teacher's Creationism Lesson Causes Stir," *Philadelphia Inquirer*, 22 August 1996, quoted in *Creation Ex Nihilo* 19 (June-August 1997): 9.

5. Phillip E. Johnson, *Defeating Darwinism by Opening Minds* (Downers Grove, Ill.: InterVarsity Press, 1997), 49–50.
6. Sue O'Brien, "Zealots Rage from Left, Too," *Denver Post,* 18 August 1996, p. F1.
7. Johnson, *Defeating Darwinism*, 51.
8. "An Affirmation of Freedom of Inquiry and Expression," National Academy of Sciences Resolution (27 April 1976), quoted in Robert V. Gentry, *Creation's Tiny Mystery* (Knoxville, Tenn.: Earth Science Associates, 1992), 7.
9. Gentry, 21–22.
10. Ibid., 170–74.
11. Henry M. Morris and John D. Morris, *Society and Creation* (Green Forest, Ark.: Master Books, 1996), 178.
12. Henry M. Morris and John D. Morris, *Science and Creation* (Green Forest, Ark.: Master Books, 1996), 213–15.
13. Ibid., 225.
14. Michael Behe, *Darwin's Black Box: The Biochemical Challenge to Evolution* (New York: The Free Press, 1996), 237.
15. Roger Lewin, "Evidence for Scientific Creationism?" *Science* 228 (17 May 1985): 837.
16. Gentry, 190–94.
17. J. Willits Lane, letter in *Physics Today* 35 (October 1982): 15, 103.
18. "A Disgrace to Biology," *Creation Ex Nihilo* 20 (December 1997–February 1998): 40.
19. Richard D. Alexander, "Evolution, Creation, and Biology Teaching," in *Evolution versus Creationism: The Public Education Controversy*, ed. J. P. Zetterberg (Phoenix: Oryx Press, 1983), 91.
20. Morris and Morris, *Society and Creation*, 163.
21. Francis Hitching, *The Neck of the Giraffe* (New York: Ticknor and Fields, 1982), 256.
22. Gale E. Christianson, *Isaac Newton and the Scientific Revolution* (New York: Oxford University Press, 1996), 73.
23. Henry M. Morris, *Men of Science—Men of God* (El Cajon, Calif.: Master Books, 1988), 25–26.
24. John Hudson Tiner, *Isaac Newton: Inventor, Scientist, and Teacher* (Milford, Mich.: Mott Media, 1975), i.
25. D. C. C. Watson, *Myths and Miracles: A New Approach to Genesis 1–11* (Acacia Ridge, Queensland, Australia: Creation Science Foundation, 1988), 112, quoted in Ann Lamont, *Twenty-One Great Scientists Who Believed the Bible* (Acacia Ridge, Queensland, Australia: Creation Science Foundation, 1995), 47.
26. Morris, *Men of Science—Men of God*, 26.
27. Ibid., 13.
28. John Hudson Tiner, *Johannes Kepler: Giant of Faith and Science* (Milford, Mich.: Mott Media, 1977), i.
29. Morris, *Men of Science—Men of God*, 13.
30. Ibid., 15.
31. Blaise Pascal, *Thoughts, Letters, and Opuscules*, trans. O. W. Wight (New York: Hurd and Houghton, 1875), 335.
32. Isaac Asimov, *Asimov's Biographical Encyclopedia of Science and Technology* (Garden City, N. Y.: Doubleday, 1972), 234.
33. Charles Coulston Gillispie, *Dictionary of Scientific Biography*, vol. 8 (New York: Charles Scribner's Sons, 1973), 380.

34. Morris, *Men of Science—Men of God*, 29–30.
35. Tryon Edwards, comp., *The New Dictionary of Thoughts: A Cyclopedia of Quotations,* rev. and enlarged by C. N. Catrevas, Jonathan Edwards, and Ralph Emerson Brown (New York: Standard Book Company, 1957), 49.
36. Gerald Holton, *Introduction to Concepts and Theories in Physical Science* (Reading, Mass.: Addison-Wesley, 1952), 166.
37. Benjamin Franklin, *The Writings of Benjamin Franklin*, vol. 10, ed. Albert Henry Smyth (New York: Macmillan, 1907), 84.
38. James Madison, *Notes of Debates in the Federal Convention of 1787* (Athens, Ohio: Ohio University Press, 1966), 209–10.
39. Christopher Columbus, *Book of Prophecies*, quoted in August J. Kling, "Columbus—A Layman 'Christ-bearer' to Uncharted Isles," *Presbyterian Layman*, October 1971, 4.
40. Ian T. Taylor, *In the Minds of Men: Darwin and the New World Order* (Minneapolis: TFE Publishing, 1991), 380.
41. Morris, *Men of Science—Men of God*, 75.
42. John Hudson Tiner, *Louis Pasteur: Founder of Modern Medicine* (Milford, Mich.: Mott Media, 1990), 90.
43. Ibid., 75.
44. *Encyclopaedia Britannica*, 15th ed., vol. 22, 503.
45. Lord Kelvin, address of 23 May 1889, quoted in Stephen Abbott Northrop, *A Cloud of Witnesses* (c. 1899; reprint, San Antonio: Mantle Ministries, 1988), 460.
46. Lamont, 184.
47. *Mathematical and Physical Papers, Lord Kelvin* (Cambridge: Cambridge University Press, 1911), quoted in Thomas G. Barnes, "Physics: A Challenge to 'Geologic Time,'" *Impact* 16 (July 1974): 1–2.
48. Morris, *Men of Science—Men of God*, 67.
49. Ibid., 68.
50. "Teaching of Organic Evolution; A Protest Meeting," *The Times* (London), 13 February 1935, p. 10.
51. Morris, *Men of Science—Men of God*, 79.
52. Richard B. Fisher, *Joseph Lister* (New York: Stein and Day, 1977), 40.
53. Rhoda Truax, *Joseph Lister: Father of Modern Surgery* (New York: Bobbs-Merrill, 1944), 121.
54. Northrop, 327.
55. Morris, *Men of Science—Men of God*, 47.
56. Jim Baggott, "The Myth of Michael Faraday," *New Scientist* 131 (21 September 1991): 44–45.
57. L. Pearce Williams, *Michael Faraday* (New York: Basic Books, 1965), 103–4.
58. Morris, *Men of Science—Men of God*, 37–38.
59. Diana Fontaine Maury Corbin, *A Life of Matthew Fontaine Maury, USN and CSN* (London: Sampson Low, Marston, Searle and Rivington, 1888), 178.
60. Morris, *Men of Science—Men of God*, 52.
61. J. G. Crowther, *Men of Science* (New York: W. W. Norton, 1936), 139.
62. Gene Adair, *George Washington Carver* (New York: Chelsea House, 1989), 54.
63. William J. Federer, *America's God and Country: Encyclopedia of Quotations* (Coppell, Tex.: Fame Publishing, 1994), 98.
64. Morris, *Men of Science—Men of God*, 63.
65. Lamont, 250.

66. Wernher von Braun, letter read by Dr. John Ford to California State Board of Education, 14 September 1972, cited in Lamont, 250.
67. Ibid.
68. Charles Ludwig, *The Wright Brothers: They Gave Us Wings* (Milford, Mich.: Mott Media, 1985), 172.
69. John Macaulay, "Sir David Brewster: Founder of the British Association," in *Short Biographies for the People*, vol. 2 (London: Religious Tract Society, 1885), 13, 15, 16.
70. Morris, *Men of Science—Men of God*, 49.
71. Ibid., 79.
72. Robert Andrews Millikan, *Evolution in Science and Religion* (New Haven: Yale University Press, 1927), 88.

Chapter Twenty-one. You and the Man Upstairs

1. Proverbs 4:3–7.
2. Isaiah 1:18.
3. G. Richard Bozarth, "The Meaning of Evolution," *American Atheist* (September 1978): 30, quoted in Duane T. Gish, *Creation Scientists Answer Their Critics* (El Cajon, Calif.: Institute for Creation Research, 1993), 371.
4. Hugh Dryden, "The Scientist in Contemporary Life," *Science* 120 (24 December 1954): 1053.
5. Isaac Asimov, "In the Game of Energy and Thermodynamics You Can't Break Even," *Smithsonian Institute Journal* (June 1970): 10, quoted in Henry Morris, "Evolution, Thermodynamics, and Entropy," *Impact* 3: 4.
6. Psalms 102:25–26.
7. Matthew 5:45.
8. Peter Kreeft and Ronald K. Tacelli, *Handbook of Christian Apologetics* (Downers Grove, Ill.: InterVarsity Press, 1994), 135.
9. Romans 6:23.
10. Nelson Glueck, *Rivers in the Desert: A History of the Negev* (New York: Grove Press, 1960), 31.
11. Irwin H. Linton, *A Lawyer Examines the Bible: A Defense of the Christian Faith* (1943; reprint, Grand Rapids, Mich.: Baker Book House, 1977), 36.
12. Ibid.
13. Simon Greenleaf, *An Examination of the Testimony of the Four Evangelists by the Rules of Evidence Administered in Courts of Justice with an Account of the Trial of Jesus*, 2nd ed. (London: A. Maxwell, 1847), 26.
14. Simon Greenleaf, letter to American Bible Society, 6 November 1852, quoted in Stephen Abbott Northrop, *A Cloud of Witnesses* (c. 1899; reprint, San Antonio: Mantle Ministries, 1988), 198.
15. Lew Wallace, "How I Came to Write 'Ben Hur,'" *Youth Companion* (2 February 1893), quoted in Northrop, 480.
16. Jeremiah 29:13.
17. Ecclesiastes 11:8, 13.
18. Matthew 6:33.
19. David R. Collins, *George Washington Carver: Man's Slave Becomes God's Scientist* (Milford, Mich.: Mott Media, 1981), iii.
20. C. S. Lewis, *Mere Christianity* (1952; reprint, New York: Touchstone, 1996), 94.
21. Psalms 30:5.

Picture Credits

1. Neg. No. 123843. Courtesy Dept. of Library Services, American Museum of Natural History
2. Neg. No. 325023. Courtesy Dept. of Library Services, American Museum of Natural History
3. Neg. No. 325097. Courtesy Dept. of Library Services, American Museum of Natural History
4. Neg. No. 325288A. Courtesy Dept. of Library Services, American Museum of Natural History
5. Neg. No. 289548. Photo by C. H. Coles. Courtesy Dept. of Library Services, American Museum of Natural History
6. Neg. No. 121929. Photo by Thane L. Bierwert. Courtesy Dept. of Library Services, American Museum of Natural History
7. Neg. No. 126175. Courtesy Dept. of Library Services, American Museum of Natural History
8. Neg. No. 2A10142. Photo by Singer. Courtesy Dept. of Library Services, American Museum of Natural History
9. Courtesy Master Books. Illustration by Earl and Bonita Snellenberger
10. Neg. No. 326675. Courtesy Dept. of Library Services, American Museum of Natural History
11. Neg. No. 271801. Courtesy Dept. of Library Services, American Museum of Natural History
12. Neg. No. 326795. Courtesy Dept. of Library Services, American Museum of Natural History
13. Neg. No. 326882. Photo by Logan. Courtesy Dept. of Library Services, American Museum of Natural History
14. AP/Wide World Photos
15. AP/Wide World Photos
16. Express Newspapers/1033/Archive Photos
17. Institute for Creation Research.
18. By permission of Johnny Hart and Creators Syndicate, Inc.
19. Neg. No. 122523. Photo by Leon Bettin. Courtesy Dept. of Library Services, American Museum of Natural History
20. Metropolitan Toronto Reference Library
21. Neg. No. 313687. Photo by H. S. Rice. Courtesy Dept. of Library Services, American Museum of Natural History
22. Neg. No. 2A17487. Courtesy Dept. of Library Services, American Museum of Natural History
23. Neg. No. 109353. Courtesy Dept. of Library Services, American Museum of Natural History
24. Neg. No. 36241. Photo by A. E. Anderson. Courtesy Dept. of Library Services, American Museum of Natural History
25. Neg. No. 336415. Photo by J. Coxe. Courtesy Dept. of Library Services, American Museum of Natural History

26. AP/Wide World Photos
27. Neg. No. 322021. Photo by Boltin. Courtesy Dept. of Library Services, American Museum of Natural History
28. Neg. No. 328862. Photo by Logan. Courtesy Dept. of Library Services, American Museum of Natural History
29. Neg. No. 313484. Photo by Julius Kirschner. Courtesy Dept. of Library Services, American Museum of Natural History
30. Neg. No. 121933. Photo by L. Bierwert. Courtesy Dept. of Library Services, American Museum of Natural History
31. Archive Photos
32. Neg. No. 37641. Photo by Kay C. Lenskjold. Courtesy Dept. of Library Services, American Museum of Natural History
33. Neg. No. 104670. Photo by Kay C. Lenskjold. Courtesy Dept. of Library Services, American Museum of Natural History
34. Popperfoto/Archive Photos
35. Neg. No. 106615. Photo by H. S. Rice. Courtesy Dept. of Library Services, American Museum of Natural History
36. Archive Photos
37. AP/Wide World Photos
38. NASA
39. Author
40. Express Newspapers/B595/Archive Photos
41. Earth Science Associates
42. Earth Science Associates
43. Neg. No. 323617. Courtesy Dept. of Library Services, American Museum of Natural History
44. Princeton Museum of Natural History
45. Stuttgart Museum fur Naturkunde; courtesy Chris McGowan, Royal Ontario Museum
46. Steven A. Austin. © Institute for Creation Research
47. Neg. No. 281234. Courtesy Dept. of Library Services, American Museum of Natural History
48. Neg. No. 14871. Photo by J. Kirschner. Courtesy Dept. of Library Services, American Museum of Natural History
49. Neg. No. 13397. Photo by J. Otis Wheelock. Courtesy Dept. of Library Services, American Museum of Natural History
50. Neg. No. 44439. Photo by Hovey. Courtesy Dept. of Library Services, American Museum of Natural History
51. Neg. No. 2A4541. Courtesy Dept. of Library Services, American Museum of Natural History
52. Steven A. Austin. © Institute for Creation Research
53. Steven A. Austin. © Institute for Creation Research
54. Steven A. Austin. © Institute for Creation Research
55. Institute for Creation Research.
56. Archive Photos
57. AP/Wide World Photos
58. Archive Photos
59. Neg. No. 315699. Photo by Charles H. Coles. Courtesy Dept. of Library Services, American Museum of Natural History

60. Neg. No. 326555. Photo by Logan-Rota. Courtesy Dept. of Library Services, American Museum of Natural History
61. Neg. No. 324398. Courtesy Dept. of Library Services, American Museum of Natural History
62. Bryan College
63. Bryan College
64. Bryan College
65. Bryan College
66. Bryan College
67. Bryan College
68. Archive Photos
69. Bryan College
70. Archive Photos
71. United Artists Corp./Archive Photos
72. Neg. No. 243016. Courtesy Dept. of Library Services, American Museum of Natural History
73. Neg. No. 312268. Photo by Irving Dutcher. Courtesy Dept. of Library Services, American Museum of Natural History
74. AP/Wide World Photos
75. Archive Photos
76. William J. Murray Evangelistic Association
77. Neg. No. 327925. Courtesy Dept. of Library Services, American Museum of Natural History
78. NASA
79. Neg. No. 125395. Courtesy Dept. of Library Services, American Museum of Natural History
80. Archive Photos
81. Neg. No. 286261. Photo by C. H. Coles. Courtesy Dept. of Library Services, American Museum of Natural History
82. Neg. No. 219467. Photo by Julius Kirschner. Courtesy Dept. of Library Services, American Museum of Natural History
83. Neg. No. 319013. Photo by Thane L. Bierwert. Courtesy Dept. of Library Services, American Museum of Natural History

Acknowledgements

I am grateful to the following publishers for permission to reproduce excerpts from their books: New Wine Press for Bill Cooper's *After the Flood*; Adler and Adler for Michael Denton's *Evolution: A Theory in Crisis*; Houghton Mifflin for Francis Hitching's *The Neck of the Giraffe*; and The Free Press for Michael J. Behe's *Darwin's Black Box.*

Thanks to Johnny Hart for letting me use the B. C. cartoon, as well as inspiring the Chicago Cubs joke. (Another of my favorites went like this—Reporter to manager: "Have any of your players ever made it to the Hall of Fame?" Manager: "Well, one of our players *almost* made it to the Hall of Fame—but when he got near Cooperstown, he switched to the wrong bus.")

I thank Dr. Emmett Williams, Dr. Robert Goette and Dr. Jonathan Sarfati for reviewing the manuscript for technical errors and providing many valuable insights.

Dr. Richard Cornelius of Bryan College checked the chapter on the Scopes trial.

Carl Hyland, Tom Eynon, Susan Scherer, Elizabeth McKinney and Paul Ingbretson also reviewed the book and provided important feedback. Paul, of course, did the outstanding cover art.

Barbara Robidoux provided very competent copy-editing as well as encouragement.

Frank Sherwin of the Institute for Creation Research and Dave Jolly of Answers in Genesis were helpful in answering many questions.

Ian Taylor, author of the superb volume *In the Minds of Men*, very graciously answered questions I had.

I thank Veronique Valdettaro for helping me translate passages from Henri Breuil's writings on Peking Man in *L'Anthropologie.*

I would like to acknowledge the following libraries, all of which contributed to the research of the book: the Mugar Memorial, Science and Engineering, Theology, and Astronomy libraries, all at Boston University; the Goldfarb Library and Gerstenzang Science Library, Brandeis University; the O'Neill Library, Boston College; the Tisch Library, Tufts University; the Hayden Library, MIT; the Ludcke Library, Lesley College; the Baker Library, Bentley College; the Goddard Library, Gordon-Conwell Theological Seminary; the library of the Episcopal Divinity School and Weston Jesuit

School of Theology; the Boston Public Library; the libraries of the Minuteman Library Network; and the New England Historic Genealogical Society.

This book occasionally poked fun at Harvard University (a tradition apparently begun by Benjamin Franklin), but all jests aside, I acknowledge my debt to this preeminent academic institution. Harvard graciously allowed me access to its unique collections, including: Widener Library, the Tozzer Science Library, the Kummel Geological Sciences Library, the library of the Museum of Comparative Zoology, and the Baker Library of the Harvard Business School.

Special thanks to the main library at the University of Arizona for supplying me with a copy of the 1890 flying reptile story from the *Tombstone Epitaph.* I also thank Richard Wheeler of Mantle Ministries for loaning me his beautiful reprint of Stephen Abbott Northrop's *Cloud of Witnesses.*

I must acknowledge the writings of C. S. Lewis, Peter Kreeft, and Ronald K. Tacelli as helping mold my understanding of theology.

Without Henry Morris's pioneering work in creation science, this book would never have been possible.

Alvart Badalian and Arrow Graphics provided very competent book and cover design.

I especially thank my wife Wei-Hsin for being so patient while I was "married" to writing this book, and for the assistance she provided in innumerable ways.

Index

Page numbers in **bold type** refer to illustrations.

Made in the USA
Middletown, DE
27 December 2018